Ecological Research at the Offshore Windfarm *a*

Federal Maritime and Hydrographic Agency

Federal Ministry for the Environment, Nature Conservation
and Nuclear Safety

Ecological Research at the Offshore Windfarm
alpha ventus

Challenges, Results and Perspectives

 Springer

Federal Maritime and Hydrographic Agency
Bernhard-Nocht-Str. 78
20359 Hamburg
Germany

Federal Ministry for the Environment,
Nature Conservation and Nuclear Safety
Stresemannstraße 128–130
10117 Berlin
Germany

This publication is part of the research project 'Accompanying ecological research at the *alpha ventus* off-shore test site for the evaluation of BSH Standard for Environmental Impact Assessment (StUKplus)' funded by the German Federal Ministry for the Environment, Nature Conservation and Nuclear Safety (grant number 0327689A). The authors assume responsibility for the content of this publication. Future activities suggested and views expressed by the authors does not necessarily reflect the policy recommendations of the publishers.

The project is part of the research initiative RAVE (Research at *alpha ventus*).

This report should be cited as:
BSH & BMU (2014). Ecological Research at the Offshore Windfarm *alpha ventus* – Challenges, Results and Perspectives. Federal Maritime and Hydrographic Agency (BSH), Federal Ministry for the Environment, Nature Conservation and Nuclear Safety (BMU). Springer Spektrum. 201 pp.
If a separate chapter is cited, the authors and the title of that chapter need to be mentioned.

Editors:
Anika Beiersdorf (anika.beiersdorf@bsh.de)
Katrin Wollny-Goerke (info@meeresmedien.de)

Library of Congress Control Number: 2014931852

ISBN 978-3-658-02461-1   ISBN 978-3-658-02462-8 (eBook)
DOI 10.1007/978-3-658-02462-8

Springer Spektrum
© Springer Fachmedien Wiesbaden 2014

Foreword

In a February 2002 strategy paper, the German government adopted the ambitious and trailblazing goal of building 20,000 to 25,000 MW of offshore wind power capacity off the German coast by 2025 to 2030. The government and all parties in the German parliament adhere to that goal in principle to this day. But merely setting goals is not enough. Attaining them takes action, and the way there is often long, hard and full of obstacles. The development of offshore wind power has involved learning the hard way, and we are far from the end of the learning curve.

Yet we embarked on the journey, committed, and confident despite all the setbacks. *alpha ventus* is a key milestone marking the breakthrough for offshore wind power in Germany. It is an important industrial and energy policy demonstration project, and numerous other offshore windfarms since built, planned or started in Germany and elsewhere in Europe have benefited from the engineering and environmental experience and expertise gained in its construction and operation.

Relying on the government's policy framework, a number of mainly mid-sized companies with experience in planning, building and operating windfarms on land set about planning windfarms at sea, in most cases between 30 and 100 km or more off the German coast. This task has not been made easier by Germany's federal structure and the resulting distribution of responsibilities, or indeed by the geography of its marine areas.

The Borkum West offshore windfarm, the pioneer project later renamed *alpha ventus*, was approved by the Federal Maritime and Hydrographic Agency (BSH) in November 2001. The German Offshore Wind Energy Foundation, which was launched in 2005, used a € 5 million grant from the German Ministry for the Environment, Nature Conservation and Nuclear Safety (BMU) to buy the rights to the windfarm in September 2005. A little over a year later, the foundation leased the rights out to the Deutsche Offshore-Testfeld und Infrastruktur GmbH (DOTI), a consortium formed by energy utilities EWE, Vattenfall and E.ON. Initial construction work began in autumn 2008.

Under the strict German regulations on marine facilities, the interplay between offshore windfarms and the marine environment was a key issue for *alpha ventus* from the outset – both during approval and in the ensuing planning, construction and operation phase. A major consideration was the impact on birdlife, marine fauna and life on the sea floor. The highest priority was and remains to minimize the impact on the natural environment.

An integral part of the approval notice for *alpha ventus* was the BSH Standard for Environmental Impact Assessment (StUK). From the start, DOTI assigned the task of meeting this standard to the German Offshore Wind Energy Foundation. Environmental assessments under the standard have three stages:

- A one-year programme before commencement of construction, to evaluate the findings and assessments on which approval is based (for *alpha ventus* this was shortened to six months given the trial nature of the windfarm and because of time constraints)
- Impacts of the construction activity on benthos, fish, marine mammals, resting birds and migrating birds; noise emissions
- Finally, a further environmental assessment – on the same topics – during a three-year operating phase, which was completed this year.

A brief anecdote illustrates the timespan covered by this major programme of study: When the first part of the research was commissioned, one of the contributors was expecting a baby. This spring, at the final briefing on results of the operating stage, I asked the husband how their child was coming along. He answered, 'Our little boy starts school this summer'. So here's to a bright future for their son – and for the ongoing expansion of offshore windfarms, which will go on providing electricity in harmony with nature and the marine environment for generations to come.

Jörg Kuhbier
Chairman, German Offshore Wind Energy Foundation

Preface

Offshore wind energy is vital in providing Germany with secure energy supplies for the long term. Fourteen years after the Federal Maritime and Hydrographic Agency (BSH) received the first application for approval of an offshore windfarm, 128 approval proceedings are now in progress for the building of offshore windfarms with some 9,500 wind turbines in the German Exclusive Economic Zone (EEZ). We have so far approved 33 windfarms with 2,250 wind turbines (as of September 2013).

Whatever the technical obstacles to building and operating windfarms far offshore, the technology has obvious advantages: Wind conditions out at sea are outstanding and subject to little turbulence, making for high and reliable performance yields. The energy is eco-friendly and incurs neither fuel costs nor carbon emission costs. No resources have to be acquired to harvest it. There is no environmentally hazardous waste to dispose of. And offshore windfarms offer low-disturbance areas where new natural habitats can evolve.

For industry, scientists and the public authorities alike, building and operating offshore windfarms beyond the twelve mile zone meant breaking new territory in terms of the engineering, scientific and legal challenges involved. While companies could make some use of experience with offshore wind energy in Denmark and the Netherlands, there was no such body of practical experience with wind turbines at depths of 40 m and distances of 30 to 100 km from the coast. Today, Germany is the industry leader and innovation driver. It is the only country in the world that builds offshore windfarms in such extreme conditions.

Offshore wind energy will only gain lasting, widespread acceptance, however, if shipping safety and protection of the marine environment are assured. In recognition of this, the Fourth National Maritime Conference on 24 and 25 January 2005 paved the way for Germany's first offshore windfarm project to be made the German test site for offshore windfarm research and development. Sited in 30 m of water some 45 km northwest of the island of Borkum, the windfarm Borkum West – subsequently renamed *alpha ventus* – thus presented the first opportunity to study the environmental impacts and gain a better understanding of marine environmental processes in and around a 'real life' offshore windfarm.

For five years, researchers and scientists accompanied the windfarm's construction and operation in a research project, 'Accompanying ecological research at the *alpha ventus* offshore test site for evaluation of the BSH Standard for Environmental Impact Assessment (StUKplus)'. The research was funded by the Ministry for the Environment, Nature Conservation and Nuclear Safety (BMU) and coordinated by BSH. Its aim was to find out more about construction and operation impacts on the marine environment, including birds, marine mammals, fish and benthic (seabed) organisms. In evaluating and analysing the project's impacts, the scientists were able to draw upon meteorological, oceanographic and ecological data collected and analysed since 2003 – before work started on the first wind turbines – at the FINO1 research platform on the periphery of the *alpha ventus* windfarm. This data made it possible to separate out impacts that specifically related to construction and operation of the windfarm.

At the *alpha ventus* test site, scientists, industry and public agencies undertook pioneering work to chart the impacts on the marine environment. This knowledge is now being incorporated in the revised BSH Standard 'Investigation of the Impacts of Offshore Wind Turbines on the Marine Environment' (StUK4). Monitoring methods during the construction and operation phase of windfarms have been adapted to offshore conditions. As the planning approval and enforcement agency for offshore plans, BSH can now require monitoring on the basis of improved scientific foundations to meet marine environmental protection needs while remaining economically viable for offshore operators.

The study findings also provide a valuable basis for further research in ecology, oceanography, geology and engineering – to the benefit of shipping, maritime technology and marine environment protection.

Monika Breuch-Moritz
President, Federal Maritime and Hydrographic Agency

Table of Contents

Table of Contents

List of Authors

Dr. Sven Adler
Swedish University of Agricultural Sciences
90183 Umeå
Sweden
Email: sven.adler@slu.se

Dennis Arreborg-Hansen
DHI
Agern Alle 5
2920 Hørsholm
Denmark
E-Mail: daha@dhigroup.com

Ralf Aumüller
Avitec Research GbR
Sachsenring 11
27711 Osterholz-Scharmbeck
Germany
E-Mail: ralf.aumueller@avitec-research.de

Anika Beiersdorf
Federal Maritime and Hydrographic Agency
Bernhard-Nocht-Str. 78
20359 Hamburg
Germany
E-Mail: anika.beiersdorf@bsh.de

Dr. Klaus Betke
itap – institute für technical and applied physics
GmbH
Marie-Curie-Str. 8
26129 Oldenburg
Germany
Email: betke@itap.de

Monika Breuch-Moritz
Federal Maritime and Hydrographic Agency
Bernhard-Nocht-Str. 78
20359 Hamburg
Germany
E-Mail: monika.breuch-moritz@bsh.de

Dr. Tim Coppack
IfAÖ – Institute of Applied Ecology GmbH
Carl-Hopp-Str. 4a
18069 Rostock
Germany
E-Mail: coppack@ifaoe.de

Christian Dahlke
Ministry for Climate Protection, Environment,
Agriculture, Nature Conservation and Consumer
Protection of the German State of North Rhine-
Westphalia
Schwannstr. 3
40476 Düsseldorf
Germany
E-Mail: christian.dahlke@mkulnv.nrw.de

Michael Dähne
Institute for Terrestrial and Aquatic Wildlife
Research
University of Veterinary Medicine Hannover,
Foundation
Werftstr. 6
25761 Büsum
Germany
E-Mail: michael.daehne@tiho-hannover.de

Hans-Peter Damian
Federal Environment Agency
Wörlitzer Platz 1
06844 Dessau-Roßlau
Germany
E-Mail: hans-peter.damian@uba.de

Dr. Jennifer Dannheim
Alfred Wegener Institute, Helmholtz Centre for
Polar and Marine Research
Am Handelshafen 12
27570 Bremerhaven
Germany
E-Mail: jennifer.dannheim@awi.de

Dr. Tobias Dittmann
IfAÖ – Institute of Applied Ecology GmbH
Carl-Hopp-Straße 4a
18069 Rostock
Germany
E-Mail: dittmann@ifaoe.de

Michael Durstewitz
Fraunhofer Institute for Wind Energy and Energy
System Technology
Königstor 59
34119 Kassel
Germany
E-Mail: michael.durstewitz@iwes.fraunhofer.de

Dr. Stefan Garthe
Research and Technology Centre
University of Kiel
Hafentörn 1
25761 Büsum
Germany
E-Mail: garthe@ftz-west.uni-kiel.de

Dr. Anita Gilles
Institute for Terrestrial and Aquatic Wildlife
Research
University of Veterinary Medicine Hannover,
Foundation
Werftstr. 6
25761 Büsum
Germany
E-Mail: anita.gilles@tiho-hannover.de

Dr. Lars Gutow
Alfred Wegener Institute, Helmholtz Centre for
Polar and Marine Research
Am Handelshafen 12
27570 Bremerhaven
Germany
E-Mail: lars.gutow@awi.de

Manuela Gusky
Alfred Wegener Institute, Helmholtz Centre for
Polar and Marine Research
Am Handelshafen 12
27570 Bremerhaven
Germany
E-Mail: manuela.gusky@awi.de

Stefan Heinänen
DHI
Agern Alle 5
2920 Hørsholm
Denmark
E-Mail: she@dhigroup.com

Katrin Hill
Avitec Research GbR
Sachsenring 11
27711 Osterholz-Scharmbeck
Germany
E-Mail: katrin.hill@avitec-research.de

Reinhold Hill
Avitec Research GbR
Sachsenring 11
27711 Osterholz-Scharmbeck
Germany
E-Mail: reinhold.hill@avitec-research.de

Wilfried Hube
Deutsche Offshore-Testfeld und Infrastruktur
GmbH & Co. KG EWE AG
Tirpitzstr. 39
26122 Oldenburg
Germany
E-Mail: kontakt@alpha-ventus.de

Annika Koch
Federal Maritime and Hydrographic Agency
Bernhard-Nocht-Str. 78
20359 Hamburg
Germany
E-Mail: annika.koch@bsh.de

Dr. Jana Kotzerka
University of Aarhus
C. F. Moellers Allé 8
8000 Aarhus
Denmark
E-Mail: kotzerka@biology.au.dk

Dr. Jochen Krause
Federal Agency for Nature Conservation
Branch Office Isle of Vilm
18581 Putbus
Germany
E-Mail: jochen.krause@bfn-vilm.de

Dr. Sören Krägefsky
Alfred Wegener Institute, Helmholtz Centre for
Polar and Marine Research
Am Handelshafen 12
27570 Bremerhaven
Germany
E-Mail: soeren.kraegefsky@awi.de

Dr. Roland Krone
Alfred Wegener Institute, Helmholtz Centre for
Polar and Marine Research
Am Handelshafen 13
27570 Bremerhaven
Germany
E-Mail: roland.krone@awi.de

Jörg Kuhbier
Becker Büttner Held
Kaiser-Wilhelm-Str. 93
20355 Hamburg
Germany
E-Mail: joerg.kuhbier@bbh-online.de

Bettina Kühn
Federal Maritime and Hydrographic Agency
Bernhard-Nocht-Str. 78
20359 Hamburg
Germany
E-Mail: bettina.kuehn@bsh.de

Dr. Christoph Kulemeyer
IfAÖ – Institute of Applied Ecology GmbH
Carl-Hopp-Str. 4a
18069 Rostock
Germany
E-Mail: kulemeyer@ifaoe.de

Florian Ladage
DHI-WASY GmbH
Office Syke
Max Planck-Str. 6
28857 Syke
Germany
E-Mail: fll@dhigroup.com

Simone van Leusen
Federal Maritime and Hydrographic Agency
Bernhard-Nocht-Str. 78
20359 Hamburg
Germany
E-Mail: simone.vanleusen@bsh.de

Dr. Klaus Lucke
IMARES Wageningen UR
PO Box 167
1790 AD Den Burg
The Netherlands
E-Mail: klaus.lucke@wur.nl

Dr. Bernhard Lange
Fraunhofer Institute for Wind Energy and Energy
System Technology
Königstor 59
34119 Kassel
Germany
E-Mail: bernhard.lange@iwes.fraunhofer.de

Dr. Bettina Mendel
Research and Technology Centre
University of Kiel
Hafentörn 1
25761 Büsum
Germany
E-Mail: mendel@ftz-west.uni-kiel.de

Thomas Merck
Federal Agency for Nature Conservation
Branch Office Isle of Vilm
18581 Putbus
Germany
E-Mail: thomas.merck@bfn-vilm.de

Eva Otto
Fraunhofer Institute for Wind Energy and Energy
System Technology
Königstor 59
34119 Kassel
Germany
E-Mail: eva.otto@iwes.fraunhofer.de

Verena Peschko
Institute for Terrestrial and Aquatic Wildlife
Research
University of Veterinary Medicine Hannover,
Foundation
Werftstr. 6
25761 Büsum
Germany
E-Mail: verena.peschko@tiho-hannover.de

Dr. Katrin Ronnenberg
Institute for Terrestrial and Aquatic Wildlife
Research
University of Veterinary Medicine Hannover,
Foundation
Werftstr. 6
25761 Büsum
Germany
E-Mail: katrin.ronnenberg@tiho-hannover.de

Bastian Schlenz
DHI-WASY GmbH
Office Syke
Max Planck-Str. 6
28857 Syke
Germany
E-Mail: bas@dhigroup.com

Dr. Andreas Schmidt
IfAÖ – Institute of Applied Ecology GmbH
Alte Dorfstr. 11
18184 Neu Broderstorf
Germany
E-Mail: a.schmidt@ifaoe.de

Anja Schneehorst
Federal Maritime and Hydrographic Agency
Bernhard-Nocht-Str. 78
20359 Hamburg
Germany
E-Mail: anja.schneehorst@bsh.de

Dr. Axel Schulz
IfAÖ – Institute of Applied Ecology GmbH
Carl-Hopp-Str. 4a
18069 Rostock
Germany
E-Mail: schulz@ifaoe.de

Henriette Schwemmer
Research and Technology Centre
University of Kiel
Hafentörn 1
25761 Büsum
Germany
E-Mail: h.schwemmer@ftz-west.uni-kiel.de

Prof. Dr. Ursula Siebert
Institute for Terrestrial and Aquatic Wildlife
Research
University of Veterinary Medicine Hannover,
Foundation
Werftstr. 6
25761 Büsum
Germany
E-Mail: ursula.siebert@tiho-hannover.de

Henrik Skov
DHI
Agern Alle 5
2920 Hørsholm
Denmark
E-Mail: hsk@dhigroup.com

Dr. Julia Sommerfeld
Research and Technology Centre
University of Kiel
Hafentörn 1
25761 Büsum
Germany
E-Mail: sommerfeld@ftz-west.uni-kiel.de

Dr. Nicole Sonntag
Research and Technology Centre
University of Kiel
Hafentörn 1
25761 Büsum
Germany
E-Mail: nksonntag@gmail.com

Katharina Teschke
Alfred Wegener Institute, Helmholtz Centre for
Polar and Marine Research
Am Handelshafen 12
27570 Bremerhaven
Germany
E-Mail: katharina.teschke@awi.de

Dr. Frank Thomsen
DHI
Agern Alle 5
2970 Hørsholm
Denmark
E-Mail: frth@dhigroup.com

Tobias Verfuß
Projektträger Jülich
Forschungszentrum Jülich GmbH
52425 Jülich
Germany
E-Mail: t.verfuss@fz-juelich.de

Ramünas Zydelis
DHI
Agern Alle 5
2920 Hørsholm
Denmark
E-Mail: rzy@dhigroup.com

Introduction

Current situation of offshore development in Germany and environmentally sound expansion of offshore wind energy

Simone van Leusen

Federal Maritime and Hydrographic Agency,
Federal Ministry for the Environment, Nature Conservation and Nuclear Safety (Eds.)
Ecological Research at the Offshore Windfarm alpha ventus,
DOI 10.1007/978-3-658-02462-8_1, © Springer Fachmedien Wiesbaden 2014

1.1   Introduction

In September 2010, the German government adopted the national Energy Concept (BMWi/BMU 2010), a roadmap for an environmentally friendly, reliable and affordable energy supply up to the year 2050. The importance of this goal increased further when in the wake of the Fukushima incident in 2011, Germany decided to phase out nuclear energy. This marked another important step in the *Energiewende* – the transformation of the energy system towards regenerative energy. The Energy Concept sets a target of increasing the renewables share of total electricity consumption to at least 35 % by 2020 and 80 % by 2050. Offshore wind energy is expected to contribute a major share of the renewable energy supply. The German government's ambitious target is for the installation of 10 GW of wind power off the German coast by 2020, rising to 25 GW by 2030.

The Federal Maritime and Hydrographic Agency (BSH) which is responsible for licensing offshore wind power projects in the German Exclusive Economic Zone (EEZ, ▢ Fig. 1.1), has so far granted 33 licences for offshore windfarms. Of these, 30 are located in the North Sea and three in the Baltic Sea (as of September 2013, ▢ Fig. 1.2). *alpha ventus*, the first ever offshore windfarm licensed and built in the German EEZ, has been in operation since April 2010.

1.2   Legal basis of offshore windfarm approval

The international legal basis for the construction of windfarms in the EEZ is the United Nations Convention on the Law of the Sea (UNCLOS). The Convention distinguishes various marine zones with different rights and duties for coastal states.

In the territorial sea, licensing procedures are the regulatory responsibility of the applicable coastal state. In the EEZ, which extends seaward beyond the territorial sea (beyond the 12 nm limit, ▢ Fig. 1.1), the coastal state has sovereign rights with regard to activities for economic exploitation and exploration of the zone, such as the production of energy from waves, currents and winds (Art. 56 para. 1a UNCLOS). Due to the geographical situation, the

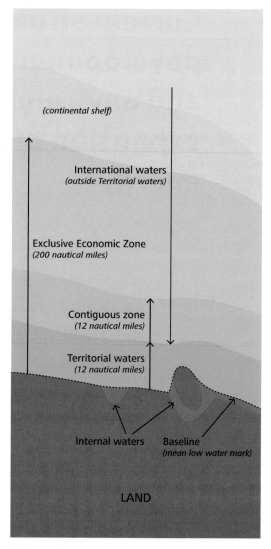

▢ **Fig. 1.1** Sea areas in international law.

German Baltic Sea EEZ is much smaller (4,500 km²) than the North Sea EEZ (28,600 km²).

In Germany, construction and operation of installations for commercial purposes in the EEZ are a federal responsibility. The Federal Maritime Responsibilities Act assigns the competency for such procedures to BSH. In the EEZ, licensing procedures are subject to a different legal regime than windfarms onshore or in the territorial sea.

The legal basis for the licensing of offshore windfarms in the EEZ is the Marine Facilities Ordinance (Seeanlagenverordnung – SeeAnlV), which first

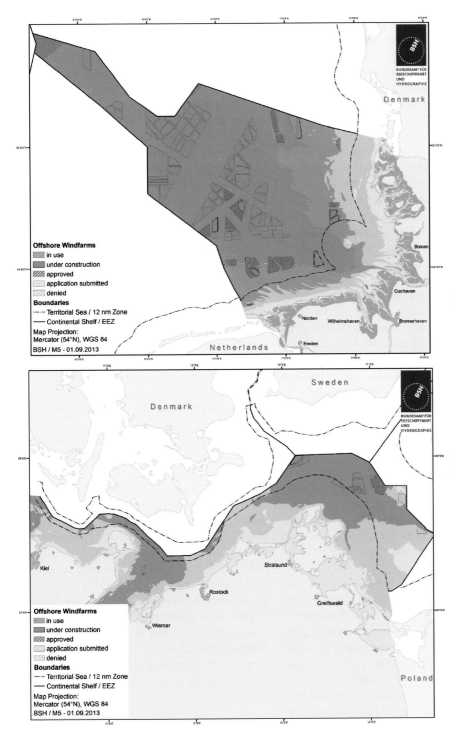

Fig. 1.2 North Sea and Baltic Sea offshore windfarm projects (as of September 2013).

came into force in 1997. The first approval granted on this basis was for the *alpha ventus* project. Other approvals followed, and experience gained in the course of these early approval procedures led to a number of amendments to the Marine Facilities Ordinance, although the general structure remained unchanged. One of the key principles of the Marine Facilities Ordinance is the freedom to apply for construction of a windfarm anywhere in the EEZ, meaning that applications are not restricted to specific areas. This distinguishes the German approach from that of neighbouring countries, where the state identifies areas in which windfarms can be built and subsequently allocates them according to the applicable procedure. BSH as the licensing authority considers during the procedure whether an approval may be granted.

Spatial plans for both the North Sea and Baltic Sea EEZ were established in 2009. The aim is to ensure sustainable development and to reconcile social and economic claims on the limited space with its ecological functions. To achieve this, priority areas and reserved areas were identified with regard to a range of uses such as shipping, exploitation of mineral resources, pipelines, submarine cables and energy generation, including wind energy. The spatial plan has to be considered in the approval process and other uses may prevent the building of offshore windfarms in certain areas. As a result, applicants for offshore windfarms began taking the spatial plans into account even before they came into force. In effect, this meant that some areas, such as shipping lanes and areas of nature conservation were excluded from consideration as potential areas for windfarm construction.

The latest spatial tool is the Offshore Grid Plan. This identifies offshore windfarms that are suitable for collective grid connections and defines the necessary cable routes and sites for connecting the windfarms to the grid. It also contains the cable routes for interconnecting cables to other European countries and descriptions of possible joint connections to help ensure system security. The goal is to provide for forward-looking and coordinated overall planning (see Chap. 2).

1.3 Marine Facilities Ordinance

The 1997 Marine Facilities Ordinance defined two major reasons to refuse an application for construction: Threats to the marine environment (as per UNCLOS, Article 1) and threats to the safety and flow of traffic. Following the first few years of experience with the new Ordinance, the grounds for refusals were refined and extended. Disruption of bird migration was added as an example of a possible threat to the marine environment, and potential threats to national or allied defence are now also now explicitly included as a reason for refusal.

A major revision of the Marine Facilities Ordinance came into force in January 2012. As a result, a planning approval procedure now needs to be carried out for projects not publicly announced before the revision. The revised Ordinance brought one key change: Additional interests (such as private interests) now have to be taken into account and the various interests weighed against each other. Threats to the marine environment, shipping safety or national or allied defence must still result in refusal of an application, however, so the effect of this change will not be as fundamental as might be thought at first sight.

An important change compared to the former version is the concentration effect. This principle as applied in planning approval procedures has the effect that licensing for offshore windfarms is combined in a single procedure and only one licence is needed. For instance, planning approval covers questions of species conservation under the Federal Nature Conservation Act which were decided on in the past by the Federal Agency for Nature Conservation (BfN) but will now be part of the BSH licence. As a special administrative authority, however, BfN will continue to be closely involved throughout the planning approval procedure as it has been in the past. Another important aim of the revised Ordinance was to prevent the 'hoarding' of licences and to tighten the requirements for prolongation of a licence.

1.4 Approval process

The approval procedure consists of several phases (▣ Fig. 1.3). Following an application, two rounds

of consultation are usually carried out to canvass other authorities, stakeholders and the public. The second round of consultation is followed by an application hearing at which the applicants have the opportunity to give a presentation on the project. Conflicting interests and uses are discussed, and the hearing sets the scope of investigations on potential impacts on the marine environment.

On the basis of the environmental monitoring, the applicant prepares an environmental impact assessment (EIA). Such an assessment is compulsory under the Environmental Impact Assessment Act (UVPG) for offshore windfarm projects comprising more than 20 turbines. Applicants are required to investigate the marine environment in the project area and predict the impact of the projected windfarm. As the first application procedures revealed a lack of experience in conducting EIAs offshore, BSH decided to draw up guidelines on the subject. Working with a large panel of experts, BSH issued the Standard for Environmental Impact Assessment (StUK4, BSH 2013) (see Chap. 5), specifying the scope of assessment and monitoring methods for the various subjects of investigation such as birds, fish and mammals. Based on the EIA and other information such as recent literature, BSH reviews whether the project poses a risk to the marine environment (including birds, fish, marine mammals, benthos, the sea bottom and marine waters).

A risk analysis dealing with the probability of vessels colliding with windfarm installations is also mandatory. Depending on the outcome, approval may be tied to additional requirements to lower collision risk (e.g. regarding observation of the area or the availability of emergency tug boats).

After receiving the documentation from the applicant, BSH passes it on to the competent authorities and organisations for comment (third consultation round). The documentation is also made available once more for public inspection and comment at BSH. The third consultation round is followed by a hearing at which comments and information concerning the marine environment, navigational safety and other interests and uses are discussed with all stakeholders. BSH then reviews whether the requirements for granting approval are met. At the same time, the applicable regional office of the Federal Waterways and Shipping Agency (WSD) verifies compliance with navigation safety and efficiency requirements.

Two applications have been turned down so far, both for environmental reasons. When approval is granted, it is published in the German Notices to Mariners (NfS), the official gazette of the Federal Ministry of Transport, Building and Urban Development (BMVBS) and in two national papers. It is available at BSH for public inspection and published on the BSH website (▶ www.bsh.de). The approval is also sent out to all authorities and organisations involved in the approval procedure.

As the impact of offshore windfarms on navigational safety and the marine environment has not yet been finally assessed, BSH generally only approves pilot-scale projects comprising a maximum of 80 wind turbines. The purpose of these smaller-scale projects is a detailed investigation of the impact of offshore windfarms on the marine environment and navigation.

In some instances, BSH has received multiple applications for the same site. The original practice was for the application that first met all requirements for approval to be decided on first. An application is deemed to meet all requirements for approval when all documents needed for decision are available to the approval authority. Under Section 3 of the revised Marine Facilities Ordinance, a certain measure of protection is now granted to the earlier applicant. This is nonetheless closely tied to a time schedule which still needs to be kept. If a

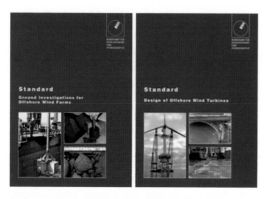

Fig. 1.4 'Standard Ground Investigations for Offshore Windfarms' (published February 2008) and 'Standard Design of Offshore Wind Turbines' (published June 2007).

project does not keep to its time schedule, it may be put on hold in favour of a competing project. The aim of this rule is to speed the development of projects while giving some protection to bona fide applicants who clearly plan to begin construction as soon as possible.

At the stage when approval is granted, projects usually do not yet have an investor to actually build the windfarm. This means a significant part of design and construction process is not carried out until a project receives approval. To accommodate and supervise construction after granting formal approval, BSH has set up a process divided into three consecutive 'clearances' that have to be obtained before work starts offshore.

With regard to site investigations and construction of offshore wind turbines, initial experience showed that, as with environmental investigations, lack of experience and standardisation posed a challenge to licenses. To improve legal and investment certainty, BSH has published a standard on the subject. Compiled in cooperation with a group of experts, the standard lays down detailed minimum requirements for mandatory geological/geophysical and geotechnical site investigations at planned windfarm sites (Standard Ground Investigations for Offshore Windfarms, BSH 2008). A further standard has been drawn up that specifies the requirements for offshore wind turbine design and ensures that all installations and structural components are certified (Standard Design of Offshore Wind Turbines, BSH 2007, ◘ Fig. 1.4).

1.5   Incidental Provisions

An important part of each approval granted by BSH for an offshore windfarm is the Incidental Provisions, which are issued in a largely standardised form. Among other things, the Incidental Provisions include a limitation of approval to a 25-year period which corresponds with the regular service life of offshore installations. An extension may be considered, however, when approval is due to expire. A standard requirement to start building before a set date aims to prevent the reservation of spaces for future use.

Other standard provisions concern the need to prepare a state-of-the-art geotechnical study and the use of best available technology in the construction of wind turbines. Foundations must be so designed as to minimise the impact of any shipping collisions. Compliance with this requirement has to be verified in an expert opinion calculating the consequences of a collision.

Construction, operation and monitoring are also subject to strict requirements regarding nature conservation. An important requirement is the use of noise mitigation methods to minimise noise emissions when installing foundations. To protect marine mammals from impulsive pile driving noise, a noise mitigation concept, submitted by windfarm operators, must ensure that the sound exposure level (SEL) does not exceed 160 dB (re 1 μPa^2s) and the peak level (L_{peak}) does not exceed 190 dB (re 1 μPa) outside of a 750 m radius of the construction site (▶ Information box *Incidental Provision 14*, see ▶ Chap. 13). Several noise mitigation measures are currently under application, investigation and testing, although none can yet be regarded as state-of-the-art (see ▶ Chap. 16). Implementation and compliance with the requirements are closely monitored by BSH and improvements have to be made if the threshold is exceeded. Daily reports must be submitted during installation, stating the work performed and any incidents.

Approvals generally require presentation of a safety plan and the installation of lights and an automatic identification system (AIS) on turbines for navigational safety. The use of environmentally compatible materials and non-glare paint is also compulsory.

After decommissioning, the offshore structures must be demolished. A bank guarantee must be provided to ensure that the costs of demolition do not fall to the taxpayer.

Considerable effort goes into ensuring the environmentally sound expansion of offshore wind energy as an important part of the national energy strategy. At the same time, experience has shown both the approval procedure and implementation to be a dynamic process. Adaptation will continue to be necessary with future technical advances and environmental research.

Literature

BMWi/BMU (2010). Energiekonzept 2050. Eine Vision für ein nachhaltiges Energiekonzept auf Basis von Energieeffizienz und 100 % erneuerbaren Energien. Erstellt vom Fachausschuss "Nachhaltiges Energiesystem 2050" des Forschungsverbunds Erneuerbare Energien. Juni 2010. 68 p.

BSH (2013). Standard Investigation of the Impacts of Offshore Wind Turbines on the Marine Environment (StUK4). Bundesamt für Seeschifffahrt und Hydrographie, Hamburg und Rostock, 86 p.

BSH (2008). Standard Ground Investigations for Offshore Windfarms. Bundesamt für Seeschifffahrt und Hydrographie, Hamburg und Rostock, 40 p.

BSH (2007). Standard Design of Offshore Wind Turbines. Bundesamt für Seeschifffahrt und Hydrographie, Hamburg und Rostock, 48 p.

The Spatial Offshore Grid Plan for the German Exclusive Economic Zone

Annika Koch

Federal Maritime and Hydrographic Agency,
Federal Ministry for the Environment, Nature Conservation and Nuclear Safety (Eds.)
Ecological Research at the Offshore Windfarm alpha ventus,
DOI 10.1007/978-3-658-02462-8_2, © Springer Fachmedien Wiesbaden 2014

2.1   Legal mandate

In 2011 the German Federal Energy Act mandated the BSH to develop and annually update a Spatial Offshore Grid Plan ('Bundesfachplan Offshore') for the German Exclusive Economic Zone (EEZ) of the North and Baltic seas. The plan is drawn up in consultation with the Federal Network Agency for Electricity, Gas, Telecommunications, Post and Railway (BNetzA), the coastal federal states and the Federal Agency for Nature Conservation (BfN). The Spatial Offshore Grid Plan (BSH 2013) defines the power cable routes and sites for the entire required grid infrastructure in the EEZ up to the 12 nm border of the territorial waters. The Grid Plan for the German EEZ in the North Sea was issued in February 2013 after two national and international consultations (■ Fig. 2.1 and ■ Fig. 2.2). The Grid Plan for the EEZ in the Baltic Sea is carried out separately.

The Grid Plan takes a sectoral planning approach and is closely linked to the Maritime Spatial Plan for the German EEZ in the North and Baltic seas (BGBL 2009a, 2009b). The plan aims to ensure coordinated and consistent spatial planning of grid infrastructure, especially for offshore windfarms in the German EEZ in the North and Baltic seas. Consent and consultation procedures with the responsible authorities are being undertaken to ensure consistency with terrestrial grid planning. The transition to territorial waters is organised by way of defining gates for the bundled routing of cables.

Pursuant to Section 17a of the Federal Energy Act, the scope of the Spatial Offshore Grid Plan covers:
- The definition of offshore windfarms suitable for collective grid connections
- Spatial routing for subsea cables required to connect offshore windfarms
- Sites for converter platforms or transformer substations
- Interconnectors
- A description of potential cross connections between grid infrastructures
- Standardised technical rules and planning principles.

The planning horizon is based primarily on German Federal Government objectives for installed offshore wind capacities for 2030.

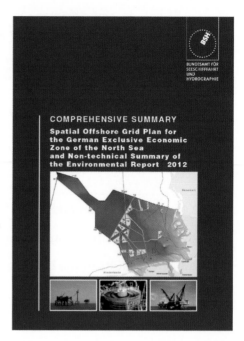

■ **Fig. 2.1** Publication of the Comprehensive Report of the Spatial Offshore Grid Plan for the North Sea EEZ 2012.

In order to issue the Grid Plan and fulfil the legal mandate, it is necessary to develop technical rules and planning principles for implementation in spatial planning. The nature of these rules and planning principles are those of principles rather than strict rules, and their stipulations should in no way hinder technological progress. Standardised technical rules pave the way for future connection of the German offshore grid to an international (North Sea) grid.

2.2   Accompanying Strategic Environmental Assessment

When drawing up the Spatial Offshore Grid Plan, an extensive Strategic Environmental Assessment (SEA) was carried out in accordance with the Environmental Impact Assessment Act, and its findings were then integrated into planning proceedings. According to Article 1 of the SEA Directive, the SEA aims to provide 'a high level of protection of the environment and to contribute to the integration of environmental considerations into the preparation and adoption of plans and programmes with a view to promoting sustainable development'.

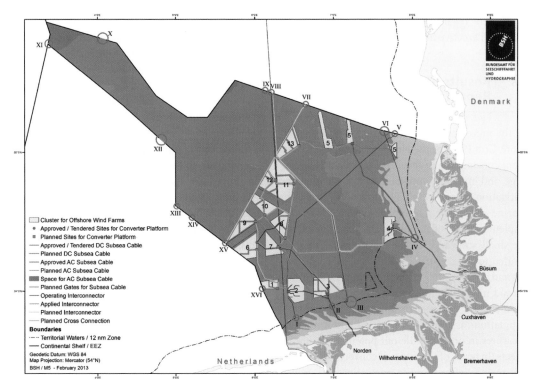

☐ **Fig. 2.2** Spatial Offshore Grid Plan for the German Exclusive Economic Zone of the North Sea 2012.

Environmental protection objectives based on international, EU und national conventions and provisions have been considered when drawing up the plan and implementing the SEA. As the findings of the SEA and the consultation process were taken into account, the stipulations of the grid plan are subject to ongoing optimisation. These have to be checked and adjusted continuously for their environmental effects. When revising the plan for the North Sea EEZ, additional planning principles were included to allow greater consideration of nature conservation concerns. In particular, a principle on noise mitigation was included in the plan.

The SEA report for the Spatial Offshore Grid Plan (BSH 2013) describes and assesses the environmental status with regard to:

- Seabed and water
- Plankton
- Benthos
- Biotope types
- Fish
- Marine mammals
- Seabirds and migratory birds

- Bats
- Biological diversity
- Landscape
- Air
- Climate
- Tangible assets, cultural heritage
- Human population and human health
- Potential interactions.

The data and information bases consist of large-scale surveys, research projects and literature studies. A vast amount of data derives from EIA studies conducted for specific offshore windfarm or grid infrastructure projects.

In the main, the SEA report focuses on the description and assessment of significant effects on the marine environment likely to arise from implementation of the plan. This relates to the planned converter platforms/transformer substations and subsea cable systems in the construction, operation and dismantling phases. The description and estimate of the state of the marine environment serve as a basis. In addition, the SEA report contains an assessment regarding species conservation and a Habitats Directive impact assessment.

2.3   Results of the Environmental Report

At the more abstract level of the SEA on the Off-shore Grid Plan for the North Sea EEZ, the assessment concludes that, according to available knowledge, and in strict compliance with prevention and mitigation measures, no significant effects on the marine environment are to be expected following implementation of the plan. The potential effects are small-scale and largely short term as they are limited to the construction phase. The coordinating and concentrating effect of the plan's stipulations should minimise the impact on the marine environment. The grid plan includes spatial and textual stipulations as regards prevention and mitigation measures in order to rule out potential significant effects. In terms of the planned converter platforms, this mainly refers to noise prevention and mitigation measures during installation of the foundations as well as environmentally-friendly lighting for the platforms. Measures to prevent and minimise any potential negative effects of subsea cables must be considered within the framework of cable route planning and technical design. This includes space-saving planning, circumvention of protected areas and the avoidance of cable crossings.

Given strict compliance with the prevention and mitigation measures, no significant effects on the protection and conservation objectives of Natura2000 sites are to be expected. Also, no significant negative effects related to prohibitions under species conservation law are associated with the converter platforms and subsea cable routes planned in the grid plan. A detailed assessment of concerns regarding species conservation and protected areas will remain subject to the individual licensing procedure. To avoid effects on protected areas, alternative solutions were examined. As part of the Habitats Directive impact assessment, alternative routes were examined for all cable routes which make use of Natura2000 network areas and for which circumvention of the protected area is possible and reasonably appropriate.

With regard to the assessment of the effects on specific nature conservation interests, in particular strictly protected habitat structures, and with regard to the cumulative consideration of bird migration, there is currently still a lack of sufficient scientific knowledge and standard evaluation methods. These effects cannot be conclusively assessed in the existing SEA and are fraught with uncertainties. A more detailed investigation must be carried out in the annual updates to the grid plan or individual licensing procedure. This also applies to the assessment regarding species conservation.

Monitoring measures are included in the SEA report in order to ascertain unforeseen negative effects. Monitoring also serves to examine the gaps in knowledge or the uncertain forecasts contained in the SEA. The monitoring results will be considered when updating the plan. The actual monitoring of potential effects on the marine environment can only begin on the specific project level, when the stipulations laid down in the plan are adhered to (effect monitoring). Thus, great significance is assigned to project-related monitoring of the effects of converter platforms and subsea cables based on BSH Standard for Environmental Impact Assessment (StUK).

Literature

BGBL (2009a). Raumordnungsplan für die deutsche ausschließliche Wirtschaftszone in der Nordsee (Textteil und Karteteil) – Anlage zur Verordnung über die Raumordnung in der deutschen ausschließlichen Wirtschaftszone in der Nordsee (AWZ Nordsee-ROV) vom 21. September 2009. Anlageband zum Bundesgesetzblatt (BGBL) Teil I Nr. 61 vom 25. September 2009.

BGBL (2009b). Raumordnungsplan für die deutsche ausschließliche Wirtschaftszone in der Ostsee (Textteil und Karteteil) – Anlage zur Verordnung über die Raumordnung in der deutschen ausschließlichen Wirtschaftszone in der Ostsee (AWZ Ostsee-ROV) vom 10. Dezember 2009. Anlageband zum Bundesgesetzblatt (BGBL) Teil I Nr. 78 vom 18. Dezember 2009.

BSH (2013). Bundesfachplan Offshore für die deutsche ausschließliche Wirtschaftszone der Nordsee 2012 und Umweltbericht. Bundesamt für Seeschifffahrt und Hydrographie (BSH), Hamburg und Rostock, February 2013, 290 pp.

The *alpha ventus* offshore test site

Wilfried Hube

Federal Maritime and Hydrographic Agency,
Federal Ministry for the Environment, Nature Conservation and Nuclear Safety (Eds.)
Ecological Research at the Offshore Windfarm alpha ventus,
DOI 10.1007/978-3-658-02462-8_3, © Springer Fachmedien Wiesbaden 2014

3.1   Introduction

alpha ventus is Germany's pioneering offshore energy project. At the time of its commissioning in April 2010, *alpha ventus* was a multiple first: The world's first far-offshore windfarm, Germany's first offshore windfarm, and the first offshore windfarm deploying 5 MW turbines. And it is one of the most successful offshore projects so far. During both 2011 and 2012, the first two full years of operation, *alpha ventus* reached a capacity factor of nearly 51 %. The submarine grid connection proved reliable too, recording not a single interruption. Each year, *alpha ventus* fed 267 GWh of power into the German grid, sufficient to supply approximately 70,000 households. It is the result of hard labour and a unique combined effort on the German high seas.

alpha ventus is a joint project. To build it, EWE AG, E.ON Climate & Renewables GmbH and Vattenfall Europe New Energy GmbH founded a joint venture in 2005, Deutsche Offshore-Testfeld- und Infrastruktur GmbH & Co. KG (DOTI). Under the name 'Offshore-Windpark Borkum-West', DOTI leased the rights under the windfarm permit from the German Offshore Wind Energy Foundation. The ownership distribution is EWE 47.5 % and E.ON and Vattenfall 26.25 % each. The total investment outlay came to €250 million, with funding of €30 million from the German Federal Ministry for the Environment, Nature Conservation and Nuclear Safety (BMU).

Twelve 5 MW class wind power turbines are in operation at the *alpha ventus* test site: Six AREVA Wind M5000 turbines and six REpower 5M turbines, built on two different types of foundations (◘ Fig. 3.1 and ◘ Fig. 3.2). The AREVA wind turbines are mounted on tripods, the REpower turbines on jacket foundations. The construction phase proper was a brief twelve months – a pioneering feat in a location with a water depth of some 30 m and at a distance of 60 km from the coast. The windfarm is run from a control centre in the town of Norden (Eastern Friesland). Its rated output is 60 MW.

3.2   Offshore challenges

The *alpha ventus* offshore windfarm is located in the open sea in Germany's Exclusive Economic Zone (EEZ), ◘ Fig. 3.3. The prevailing wind at the site is 210–240° (south-westerly) with an average wind speed of 10 m/s, equivalent to Beaufort force 5. In heavy weather, the swell can produce waves more than 10 m high, with an average wave height of 6 to 8 m. The main direction of swell is 330° (north-westerly).

The location far offshore reflects the specific conditions of the German North Sea. Planning permission is only granted for offshore windfarms if they are located beyond the Wadden Sea World Natural Heritage site and off the shipping lanes of the German Bight traffic separation schemes.

On the other hand, a location in the open North Sea guarantees excellent wind yield. The original forecasts of 3,900 full load hours per annum compare with approximately 2,000 hours for land-based sites. About 4,450 full load hours per year were registered during 2011 and 2012, far in excess of forecast. At the same time, water depths of up to 40 m, aggressive salt-laden air, strong and often gusty winds and the swell together place extreme demands on installation logistics, construction, operation and maintenance. The challenges created by the environmental conditions also force up the costs of investment and operation compared with near-coastal offshore locations and onshore windfarms.

The wind turbines are arranged in a grid formation with a mutual separation of roughly 800 m. Four rows of three wind turbines each mark out a rectangle covering a total area of 4 km², equivalent to around 500 soccer fields. All general shipping and fishing vessels are prohibited from entering the entire area of the windfarm.

The offshore structures do not rest directly on the seabed. The phenomenon of scour holing – where sediment transportation driven by the interaction of waves and currents forms hollows around embedded structures – resulted in a foundation design isolated from the sea bed. The foundations actually stand on pillars embedded up to 35 m deep in the sea floor, to which they are permanently fixed. The foundations are also designed to withstand 'freak' waves. The submarine cables are dug into the sea floor as well to avoid damage, for example from anchors.

Above sea level, all key technical components are housed in air-conditioned compartments to

AREVA Multibrid M5000   REpower 5M

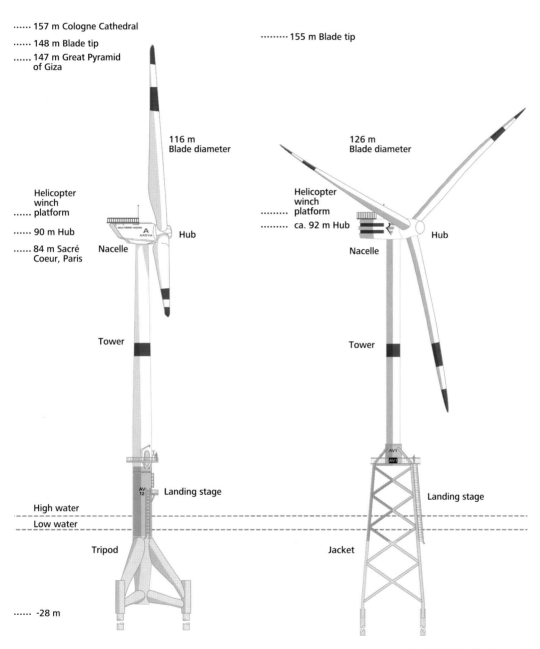

AREVA Multibrid M5000

······ 157 m Cologne Cathedral
······ 148 m Blade tip
······ 147 m Great Pyramid
        of Giza

116 m
Blade diameter

Helicopter
winch
······ platform
······ 90 m Hub
······ 84 m Sacré
        Coeur, Paris

Hub
Nacelle

Tower

Landing stage
High water
Low water

Tripod

······ -28 m

REpower 5M

········· 155 m Blade tip

126 m
Blade diameter

Helicopter
winch
········· platform
········· ca. 92 m Hub

Hub

Nacelle

Tower

Landing stage

Jacket

As of 04/2009, [Not to scale]

◘ **Fig. 3.1** AREVA wind turbines mounted on tripods; REpower wind turbines mounted on jackets.

Transformer Station

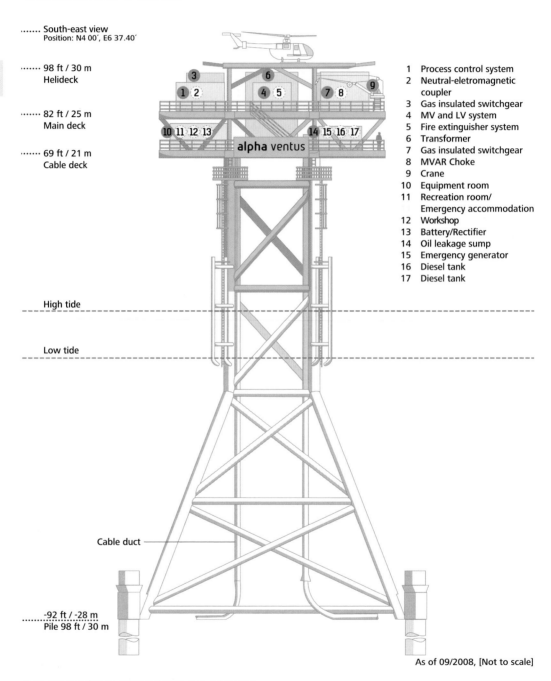

······· South-east view
     Position: N4 00′, E6 37.40′

······· 98 ft / 30 m
     Helideck

······· 82 ft / 25 m
     Main deck

······· 69 ft / 21 m
     Cable deck

High tide

Low tide

Cable duct

-92 ft / -28 m
Pile 98 ft / 30 m

alpha ventus

1 Process control system
2 Neutral-eletromagnetic
    coupler
3 Gas insulated switchgear
4 MV and LV system
5 Fire extinguisher system
6 Transformer
7 Gas insulated switchgear
8 MVAR Choke
9 Crane
10 Equipment room
11 Recreation room/
     Emergency accommodation
12 Workshop
13 Battery/Rectifier
14 Oil leakage sump
15 Emergency generator
16 Diesel tank
17 Diesel tank

As of 09/2008, [Not to scale]

■ **Fig. 3.2** Transformer station (substation) at *alpha ventus*.

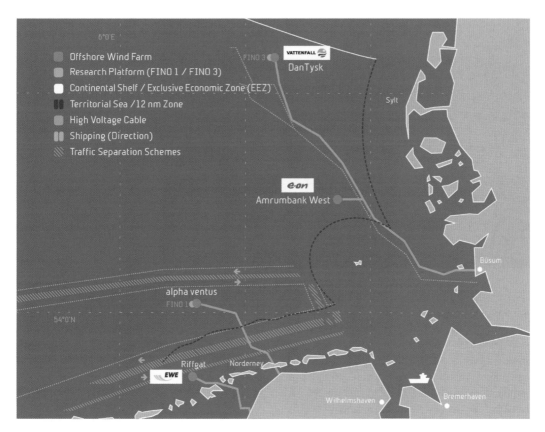

□ Fig. 3.3 Location of the *alpha ventus* offshore test site.

avoid the corrosive influence of the sea air. Most technical components are also duplicated to minimise downtime for turbines and the windfarm as a whole. All structures are accessible both by boat and helicopter. The substation has a landing pad and each turbine has a winching area on top of the nacelle.

3.3 Operation of the offshore windfarm

For the most part, *alpha ventus* is operated from land. All wind turbines and the transformer station are monitored and controlled from the control centre in Norden. It is nonetheless necessary for personnel to go offshore on a regular basis. This includes routine servicing and maintenance work as well as ongoing testing of the twelve wind turbines. The testing takes place on a four-year cycle and cov-

□ Fig. 3.4 During construction phase, shortly before a rotor is mounted (photo: BMU / webcam FINO1, October 2009).

◨ **Fig. 3.5** *alpha ventus*: wind turbines and transformer station (photo: DOTI / Matthias Ibeler 2011).

ers everything from the base of the foundations to the tips of the rotor blades.

Work on the high seas is subject to stringent safety rules. The wind turbines and the transformer station are each fitted with comprehensive safety equipment. This includes full first aid kits and communications including several telephones on each wind turbine. Service personnel must have certified training on all aspects of safety at sea and for helicopter flights, and are given regular occupational health checks. Protection and safety drills are based on strict, regulator-sanctioned procedures and routines including full logging of communications with the control centre. This helps ensure that all work is coordinated and also enables a quick response to any sudden changes in weather conditions. The safety measures are augmented by emergency plans that are fully approved by the public maritime traffic monitoring organisations and sea rescue services.

3.4   The control centre in Norden

The town and port of Norden/Norddeich in the district of Aurich is the onshore headquarters of *alpha ventus*. The control centre operates in shifts, normally with two supervisors per shift. All information and data are gathered there. The windfarm operating status is displayed in real time on banks of monitors showing images, maps, and data. The

collected operating data includes wind speeds, output levels, oil temperatures and nacelle alignments. The data is monitored and evaluated in a condition monitoring system (CMS) to allow early identification of unusual values so action can be taken as needed. The shift supervisors coordinate and monitor the work of the service teams at the windfarm and are also available at all times as a point of contact on the mainland. A number of adjustable and fixed webcams are installed at the windfarm to allow shift supervisors to also monitor helicopter flights and shipping movements.

3.5   Offshore power transmission

Within the windfarm, power is transmitted from wind turbines to the offshore transformer station along 33 kV sea cables. Some 16 km of cable are buried at least 60 cm deep in the sea bed. The offshore transformer station steps up the voltage to 110 kV and power is finally transmitted to the mainland along an approximately 60 km sea cable that also crosses the island of Norderney. On the mainland, the electricity is fed into the Hagermarsch substation from where it enters the German grid. The grid operator, TenneT TSO GmbH, is responsible for operating the offshore grid feed-in. The sea cable also includes integrated optical fibre data lines to provide modern communication and monitoring system links to the windfarm.

3.6 Retrospective: The building period

alpha ventus started life as the Borkum West offshore test site and was the first offshore windfarm for which the Federal Maritime and Hydrographic Agency (BSH) granted official planning permission. The German Offshore Wind Energy Foundation leased the rights in 2005 with financial support from the BMU and subsequently leased them on to the *alpha ventus* DOTI consortium. The Foundation has accompanied the project as licence holder to the present day.

Autumn 2008 saw the start of initial preparatory work. The first structure to be constructed was the offshore transformer station at the south-eastern corner of the future windfarm. This was followed by the laying of the 60 km sea cable connecting the windfarm to the German national grid. Cable laying was completed in spring 2009. In April 2009, work started on constructing the wind turbines. The first milestone was finally reached on 1 June that year with the anchoring of the six tripod foundations for the AREVA wind turbines. This was followed by step-by-step assembly of the first tower segments. From mid-July onwards, the turbines were completed in sequence with the top tower segment, the nacelle and the rotor star, followed by commissioning. Work on the foundations for the six REpower units commenced in June 2009. In September 2009, the jackets – the basic foundation structures for the REpower turbines – were transported to the building site and anchored into place on the seafloor in a short six-day period. The first REpower 5M was completed by 30 September 2009 (■ Fig. 3.4). On 16 November 2009, *alpha ventus* was completed in its entirety, with the twelfth turbine finally assembled in the early hours of that morning. This was followed by commissioning of the wind turbines, which was completed in April 2010 (■ Fig. 3.5). The actual offshore building period for all twelve wind turbines therefore spanned seven months, from April to November 2009.

■ **Fig. 3.6** Wilfried Hube, managing director of DOTI, on board the construction vessel 'Thialf' (photo: DOTI / Matthias Ibeler 2009).

3.7 Outlook: *alpha ventus* as a reference project

alpha ventus can be regarded both as a stress test and a pilot project for a new generation of wind turbines in the open sea. The objective was to demonstrate the technical and technological viability as well as the economic feasibility of offshore technologies, including far-offshore locations. The knowledge and experience gained in the process relate not only to the technology but also to the process as a whole: From approval procedures and environmental audits via the safety concept, the tendering process, the logistics and the various construction stages right through to individual maintenance tasks and detailed operating procedures. The experience gained in the operation of the windfarm represents crucial basic knowledge for Germany's nascent offshore wind industry. The DOTI operating companies, EWE, E.ON and Vattenfall, have already put the experience gained in the construction and operation of *alpha ventus* to good use for future projects. Construction of the Riffgat (EWE), Amrumbank (E.ON) and Dan Tysk (Vattenfall) offshore windfarms started in 2012 and 2013. Riffgat was successfully completed in August 2013 (■ Fig. 3.3).

The *alpha ventus* offshore test site: Impressions of the construction phase

◘ Jacket foundation (photo: DOTI / Matthias Ibeler, DOTI / Jan Oelker).

◘ Installation of the rotor (photo: DOTI / Wolfgang Scheer).

◘ Tripod foundation (photos:
DOTI / Jan Oelker).

The RAVE research initiative: A successful collaborative research, development and demonstration programme

Eva Otto, Michael Durstewitz, Bernhard Lange

Federal Maritime and Hydrographic Agency,
Federal Ministry for the Environment, Nature Conservation and Nuclear Safety (Eds.)
Ecological Research at the Offshore Windfarm alpha ventus,
DOI 10.1007/978-3-658-02462-8_4, © Springer Fachmedien Wiesbaden 2014

4.1   Introduction

In 2002, the German government set the target of reaching 20 to 25 GW of offshore wind power capacity installed in German waters by 2030 (BMU 2002). This planned rapid expansion was mainly driven by political aims, but was soon supported by a range of stakeholders, including the offshore wind energy industry. However, developing a new technology, a completely novel industrial sector and an innovative and independent research discipline takes more than just a wish. In 2007, when the first offshore windfarm was yet to be built in Germany, the government decided both to involve more stakeholders and to provide substantial funding. With financial support from the Federal Ministry for the Environment, Nature Conservation and Nuclear Safety (BMU), a trans-disciplinary consortium was established comprising representatives from science, industry and administration. This community is known as the RAVE research initiative. RAVE (Research at Alpha VEntus) was set up to accompany the construction and operation of alpha ventus, Germany's first offshore windfarm, with a range of research, development and demonstration activities. alpha ventus is the first big offshore wind research and demonstration site worldwide, and has since become a highly impressive research focus. The initial objectives of RAVE were:

- To find solutions for a number of challenging issues in relation to the utilisation of offshore wind power in Germany
- To promote the use of and further develop the technology for wind power at sea
- To create a national offshore wind research network (BMU 2002, 2006).

4.2   Actors and organisation

Over 30 individual projects conducted by over 50 universities, research organisations and companies have now been carried out under the RAVE umbrella. More will be integrated into the consortium over time. A range of large-scale, joint RAVE projects have fostered both interdisciplinary collaboration and knowledge acquisition and transfer. The RAVE consortium provides an additional trans-disciplinary framework, resulting in a stunning pool of offshore wind power expertise. Scientists and managers actively and continuously provide solutions and new know-how, thus contributing to attainment of the RAVE targets.

The heart of the RAVE consortium is the steering committee, which orchestrates common activities overall and brings together representatives from the joint projects, the test site operator, the two turbine manufacturers, the administrative supervisor, Project Management Jülich (PtJ), and Fraunhofer Institute for Wind Energy and Energy System Technology (IWES), which coordinates RAVE. Around 25 professors, scientists, project leaders, decision makers and subcontractors meet regularly to inform each other about the status of specific projects, achievements made, bottlenecks and delays, and to decide about current strategic issues concerning the RAVE initiative. The steering committee has proven to be a successful instrument in exchanging scientific and operational experience, and serves as a starting point for new collaboration both within the RAVE community and with external partners. The roles of the operator and the manufacturers in the consortium are of great importance. As owners of the turbines, the windfarm and the data, they support the measurement applications at the turbines. Additionally, as the tenant of the test site and as a major promoter of offshore wind power development in Germany, the German Offshore Wind Energy Foundation enabled the realisation of the test site. A big consortium such as RAVE clearly needs an overall coordination team. The coordination project, led by Fraunhofer IWES in Kassel, aims to link the RAVE projects, make use of synergies and communicate the initiative to the general public.

4.3   Measurements and data

For the RAVE research projects – primarily applied and demonstration projects – comprehensive measuring data is a must. A huge measurement programme has been set up at sea to conduct state-of-the-art science. All measurements required for the research projects are performed by a central measurement service led by the Federal Maritime

Fig. 4.1 Sketch of the REpower 5M (AV4) offshore wind turbine at the *alpha ventus* test site. Markings on the structure and in the water indicate locations and type of instrumentation used for RAVE measurements.

- strain gauges
- acceleration
- acoustic sensors
- hydrographic sensors
- sonars
- corrosion
- wind (USA, LiDAR)
- SCADA
- video cam, radar

and Hydrographic Agency (BSH). The project coordinates data demand, and advises project managers regarding their sensor, sample and measurement needs. It also plans, installs, operates and maintains measurement equipment as a service for all RAVE partners. Data is first validated by the responsible data supervisors: BSH for environmental measurements, DEWI GmbH for AREVA Wind turbines, GL Garrad Hassan (GL) for REpower Systems turbines, the turbine manufacturers for data from their machines, and OFFIS, which is in charge of development and operation of the RAVE data warehouse. Following this validation, data is then uploaded to the RAVE research archive. In this way the data is made available online for accredited RAVE researchers. The measurements carried out include load conditions, operational and installation noise, along with meteorological, oceanographic and geological parameters. This has resulted in an invaluable and unique measurement dataset: Four turbines are extensively instrumented and further sensors are installed in the surrounding waters, in the foundations and at offshore and onshore substations. Altogether, more than 1,200 sensors have been installed (Fig. 4.1), generating over 14 TB of stored data so far (as of June 2013).

4.4   Research focus

Since the start of the research initiative in 2007, cost reduction through activities such as technical

Funding & project supervision
BMU – political initiative and project funding; PtJ – administrative controlling

Coordination of the RAVE initiative
Fraunhofer Institute for Wind Energy and Energy System Technology IWES, Kassel

Measurements, data management, research archive
Federal Maritime and Hydrographic Agency BSH; DEWI; GL Harrad Hassan; OFFIS

Owner, operator, manufacturer

Offshore Foundation – owner of the rights

DOTI – owner of the turbines and operator, wind farm management

AREVA Wind & REpower Systems – manufacturers

RAVE – REpower Components

REpower Systems SE

RAVE – OWEA

ForWind – University of Oldenburg

RAVE – AREVA Wind M5000 Improvement

AREVA Wind GmbH

RAVE – Offshore WMEP

Fraunhofer IWES

RAVE – LIDAR

ForWind – University of Oldenburg

RAVE – LIDAR II

ForWind – University of Oldenburg

RAVE – REpower Blades

REpower Systems SE

RAVE – GW Wakes Teilprojekt A

ForWind – University of Oldenburg

RAVE – OWEA Loads

University of Stuttgart

RAVE – TUFFO

Karlsruhe Institute of Technology (KIT)

RAVE – Operational Noise

Flensburg University of Applied Sciences

RAVE – Ecology

BSH – Federal Maritime and Hydrographic Agency

RAVE – Acceptance

Martin-Luther-University Halle-Wittenberg

RAVE – Hydro Sound

Leibniz Universität Hannover – ForWind

RAVE – Sonar Transponder

Leibniz Universität Hannover – ForWind

RAVE – Geology / Oceanography

BSH – Federal Maritime and Hydrographic Agency

RAVE – UFO

Hochschule Bremerhaven fk-wind

RAVE – Foundations

BAM – Federal Institute for Materials Research and Testing

RAVE – GIGAWIND alpha ventus

Leibniz Universität Hannover – ForWind

RAVE – Foundations Plus

BAM - Federal Institute for Materials Research and Testing

RAVE – Grid Integration

Fraunhofer IWES

Turbine Technology and Monitoring

Foundation and Support Structures

Environment

Grid Integration

◘ Fig. 4.2 The RAVE research initiative at a glance.

optimisation of wind energy turbines, innovative control techniques and improvements to foundation structures have been a key research focus. A second focus is ecological and environmental research. RAVE projects are clustered around five themes: Foundation and support structures, turbine technology and monitoring, environment (including health and safety), grid integration, and cross-sectional projects. ◘ Figure 4.2 gives a brief overview of the RAVE research projects and the coordinating institutions for the joint projects (not all project leaders are mentioned). Environmental research accounts for the second largest number of projects, thus highlighting the significance of this research area.

4.5   Achievements

Summarising the initiative's achievements, it can be said that RAVE has been extremely successful. *alpha ventus* delivered outstanding production and availability figures in its first years of operation: 267 GWh of electricity was fed into the grid, both in 2011 and 2012, corresponding to 4,463 full load hours. A a new class of turbines was used to demonstrate reliable operation under extreme offshore conditions: In 2011, turbine availability was up to 97 %. The largest and most comprehensive measurement programme at an offshore windfarm worldwide has delivered unprecedented industrial-scale in-situ data for use in model development and validation. RAVE research projects have delivered significant results based on that data. To date, about 500 scientific papers, posters and talks have resulted from such projects. Information about the individual projects is available together with research results and references on the RAVE portal, ► www.rave-offshore.de. A RAVE science documentary can be viewed at ► www.youtube.com/user/RAVEoffshore/feed. Also, over 300 conference delegates attended the RAVE International Conference 2012, making the conference a huge success (► www.rave2012.de).

An extraordinary result of RAVE is that it has given rise to a dedicated community that is now capable of supporting the offshore wind industry in tackling the challenges ahead. The RAVE partners have collaborated closely now for over six years, assisting one another in various tasks. By continuing this close collaboration between the participating companies and research organisations, especially when it comes to measuring activities and data management, the RAVE success story will go on.

Literature

Federal Ministry for the Environment, Nature Conservation and Nuclear Safety (BMU) (2002). Strategy of the German Government on the use of offshore wind energy in the context of its national sustainability strategy.

Federal Ministry for the Environment, Nature Conservation and Nuclear Safety (BMU) (2006). Gemeinsame Erklärung zur Errichtung des Offshore-Testfeldes und zur Initialzündung für die deutsche Offshore-Windenergieentwicklung.

Accompanying ecological research at *alpha ventus*: The StUKplus research project

Anika Beiersdorf

Federal Maritime and Hydrographic Agency,
Federal Ministry for the Environment, Nature Conservation and Nuclear Safety (Eds.)
Ecological Research at the Offshore Windfarm alpha ventus,
DOI 10.1007/978-3-658-02462-8_5, © Springer Fachmedien Wiesbaden 2014

5.1   Environmentally compatible expansion of offshore wind energy

When the first applications for offshore windfarm projects were submitted to the Federal Maritime and Hydrographic Agency (BSH) many questions concerning possible impacts were far from being answered satisfactory. Although some practical experience had been gained in other European countries, concrete data was difficult to obtain and difficult to apply to German projects which were located up to 150 km offshore and at water depths of up to 50 m. With more approvals being granted and the first windfarms under construction, cumulative evaluation of possible impacts became more and more important.

Development of the *alpha ventus* offshore windfarm was initiated in 2005. As the nucleus for Germany's offshore wind energy sector, the test site is a focal point for technical, ecological and social research under the umbrella of the RAVE initiative. Ecological research at *alpha ventus* is brought together in the StUKplus research project, funded by the Federal Ministry for the Environment, Nature Conservation and Nuclear Safety (BMU), ◘ Fig. 5.1. This project supplements the mandatory ecological monitoring by windfarm operators according to BSH Standard for Environmental Impact Assessment (StUK) (▶ Information box: *Standard for Environmental Impact Assessment*). The StUKplus project sets a wider frame in size, scope and content than 'ordinary' monitoring. In 2008, BSH was put in charge of coordinating the StUKplus project over a time period of almost six years – from May 2008 to March 2014. StUKplus is the most important German research project on potential environmental impacts related to offshore windfarms so far.

During the extensive field research programme, novel observation methods and technologies such as aerial digital survey techniques and new bird migration radars were applied for the first time in German waters. The purpose of the ecological research was to gain a better understanding of possible environmental impacts of offshore windfarms and to evaluate the second update of the StUK standard (StUK3, BSH 2007) which was used for the first time in an offshore windfarm during construction

◘ **Fig. 5.1** Logo of the StUKplus research project.

and operation – i. e. in the *alpha ventus* offshore test site.

The main research topics aimed to provide answers to the following questions:

How do habitats change for benthic organisms and fish close to the foundations? How are these organisms influenced by the artificial reef structures? How do habitats change as a result of fisheries being excluded from the windfarm area?

How do birds react to the rotating, illuminated wind turbines? Is there a risk of migratory birds colliding with the turbines at sea? Will resting birds avoid the windfarm area?

What impacts will noise-intensive construction work have on marine mammals? Will they continue to use the windfarm area as habitat and how can they be protected from noise? How do they react to operating noise?

Standard for Environmental Impact Assessment: Investigation of the Impacts of Offshore Wind Turbines on the Marine Environment (StUK)
One of the German government's aims is to ensure the environmentally compatible development of offshore wind energy. Offshore windfarms can consequently only be approved if major harmful effects on the marine environment are sufficiently unlikely. A central

component of the approval process for offshore windfarms is therefore an environmental impact assessment (EIA), which forecasts the extent to which construction of a windfarm will endanger the marine environment. An EIA is accordingly stipulated in German windfarm permits. The EIA follows a standardised design laid down by the approving government agency, the BSH. The requirements are described in detail in BSH Standard for Environmental Impact Assessment (StUK4, BSH 2013), ■ Fig. 5.2. The aim is to identify environmental impacts at the earliest stage possible in order to avoid and minimise potential effects on marine organisms.

The StUK standard lays down standardised monitoring requirements covering:
- Benthos
- Fish
- Resting birds
- Migratory birds
- Marine mammals (including guidelines for underwater noise measurements)
- Bats
- Landscape

The standard provides technical details of the investigation and monitoring surveys to be carried out to protect species of conservation interest in German offshore areas. The objects and scope of investigations, the methods to be applied and the presentation of results are described to create a consistent evaluation basis for all offshore monitoring studies. The monitoring results produced in the offshore windfarm projects are thus quality proofed, mutually comparable, and provide detailed cumulative insights at a large temporal and spatial scale for much of the German EEZ. Compliance with the standard is obligatory for applicants and permit holders during planning, construction and operation of a windfarm.

StUK uses a Before-After-Control-Impact (BACI) design. This involves sampling in an area planned to be affected by a development and in a reference (control) area not affected by the development. Each area is sampled before, during and after the potential disturbance. The StUK scope therefore specifies two years of baseline monitoring prior to construction, monitoring during the entire construction phase, and three years of monitoring during operation of the windfarm. The monitoring surveys have to be closely coordinated with BSH and are supervised throughout. The raw data from the EIA studies has to be handed to BSH for quality assurance and becomes part of a large database of biotic information.

StUK was implemented in December 2001 for the first offshore windfarm applications and has been continuously updated since. The current version is the third update published in October 2013. Compilation and evaluation of StUK is coordinated by BSH and based on an extensive consultation process with experts from research institutions, consultancies and licensing authorities operating in German offshore areas. StUK is consequently broadly accepted by research and policy institutions, environmental organisations (NGOs) and industry, i. e. windfarm operators.

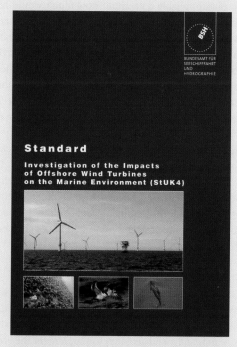

■ **Fig. 5.2** Standard for Environmental Impact Assessments (StUK4, published October 2013) issued by BSH.

5.2   Identifying environmental risks at the earliest possible stage

The StUKplus research project aimed to predict the extent to which future windfarms will endanger the maritime environment and to identify environmental impacts at the earliest possible stage in order to avoid adverse effects on marine organisms. The StUKplus project followed a coordinated, synergetic approach: The mandatory StUK3 monitoring was supplemented with additional research projects conducted over a larger area, at higher intensity with new investigation methods covering benthic organisms, fish, passage migrants, migratory birds and marine mammals. The studies were supplemented by underwater noise measurements. Being designed as effect monitoring, a before/after comparison study was carried out. Comprehensive preliminary studies were conducted at the test site before construction as early as 2008. A large database was compiled, and used in assessing the likely impact of the construction and operational phases. To obtain reliable data about how fauna would react to the windfarm in their habitats, long-term studies were performed extending several years into the operating phase. BSH's large marine environmental database of EIA studies was combined with the StUKplus research data and jointly evaluated. The StUKplus research partners spent several years studying the environmental impact of offshore windfarms (◘ Table 5.1). Their know-how and experience were applied to a real-life project, the *alpha ventus* offshore test site.

The various research projects presented in this book are briefly described below. Detailed results of the ecological monitoring and the research accompanying the project is presented in the chapters that follow.

Chapter 8: *Research on oceanography and geology*
The RAVE Service Project, coordinated by BSH, is part of the RAVE research initiative and surveyed abiotic conditions at the *alpha ventus* test site (oceanography and geology) to supplement the ecological information collected by the biologists with extensive abiotic data. The focus of the geological investigation programme was on interactions between the piled foundations, the seabed and oceanographic parameters (e.g. wave, tide and current) and sediment shifting in the windfarm. The scour depths at the foundations and their development in the test site area were measured by single-beam echosounders and multibeam sonar. Numerous sensors were mounted on wind turbines for detailed investigation of the oceanographic conditions. Temperature sensors, directional wave riders, video cameras and current sensors captured data to analyse interactions between the windfarm and the ocean.

Chapter 9: *Research on benthos and mobile megafauna*
Benthos-related research at *alpha ventus* was carried out by the Alfred Wegener Institute (AWI), Helmholtz Centre for Polar and Marine Research. Benthic fauna were surveyed within the windfarm area and in a reference area between 2008 and 2011. Soft-bottom epifauna – mostly mobile, invertebrate organisms such as crabs, shrimps and starfish that live on top of the seafloor – were sampled with a beam trawl. Soft-bottom infauna – animals such as clams, tubeworms and burrowing crabs that live inside the sediment – were sampled using a van Veen bottom grab. Assemblages of fouling organisms such as mussels and sea anemones colonising turbine foundations and mobile demersal megafauna were sampled by scientific divers (◘ Fig. 5.4). The mobile demersal megafauna were also surveyed visually along belt transects extending on the seafloor away from the turbine foundations.

Chapter 10: *Research on fish*
A study on fish was conducted by the Alfred Wegener Institute (AWI), Helmholtz Centre for Polar and Marine Research, covering the pre-construction phase, the construction period and the first years of the operational phase at *alpha ventus*. Multi-day ship-based hydroacoustic surveys were carried out in spring, summer and autumn inside and outside the windfarm. The measurements were supplemented by a recently developed stationary long-term hydroacoustic measurement system. Additional net catches with a pelagic trawl allowed the fish species composition and size distribution to be identified. The stomach contents of horse mackerel and mackerel were analysed to examine the influence of the windfarm on the feeding behaviour of pelagic fish.

◻ **Table 5.1** Overview of the StUKplus research projects.

Sub-project	Duration	Research partner
Investigation of the impacts of wind turbines on fish and mobile megafauna in the *alpha ventus* test site	01.07.2008–31.08.2012	Alfred Wegener Institute (AWI), Helmholtz Centre for Polar and Marine Research
Joint evaluation of research data, data from monitoring programmes and EIA studies as a holistic approach to ecological effect monitoring in the *alpha ventus* test site	01.09.2008–30.04.2012	Alfred Wegener Institute (AWI), Helmholtz Centre for Polar and Marine Research
Completion of time series during the operational phase and assessment of benthic changes in extended site-specific effect monitoring	01.10.2008–31.08.2012	Alfred Wegener Institute (AWI), Helmholtz Centre for Polar and Marine Research
Test site research on bird migration in the area of the *alpha ventus* test site	01.07.2008–31.08.2013	Avitec Research GbR
Evaluation of bird migration data recorded continuously at the FINO1 research platform (2008–2012).	01.08.2009–31.08.2013	Avitec Research GbR
Assessment of collision risk of migratory birds using the VARS camera system	01.10.2008–31.08.2013	Institute of Applied Ecology GmbH (IfAÖ)
Monitoring of evasive movements of migratory birds using the Bird Scan method (fixed pencil beam radar)	01.10.2008–31.08.2013	Institute of Applied Ecology GmbH (IfAÖ)
Joint evaluation of seabird data for ecological effect monitoring in the *alpha ventus* test site	01.06.2008–30.09.2013	Research and Technology Centre (FTZ)
Studies on possible habitat loss and behavioural changes in seabirds in the *alpha ventus* offshore test site	01.10.2009–30.09.2013	Research and Technology Centre (FTZ)
Studies on the impacts of construction and operation of the *alpha ventus* offshore test site on marine mammals	01.06.2008–30.09.2013	University of Veterinary Medicine Hannover, Foundation (TiHo)
Joint evaluation of marine mammal data for ecological effect monitoring in the *alpha ventus* test site	01.06.2008–31.08.2012	University of Veterinary Medicine Hannover, Foundation (TiHo)
Long-term data analysis and modelling of harbour porpoise distribution in the *alpha ventus* test site as input to a decision support tool for marine spatial planning	01.01.2013–30.09.2013	DHI / DHI-WASY GmbH
Measurement of pile driving and operational noise at larger distances from the *alpha ventus* test site and model-based processing	01.07.2008–31.08.2011	Institute for Technical and Applied Physics GmbH (itap)
Underwater noise in offshore windfarms: Harmonisation of definitions, methods and evaluation with regard to needs-oriented parameters	01.10.2010–30.11.2011	Müller-BBM GmbH

Chapter 11: *Research on resting birds*

Comprehensive studies on resting birds were conducted by the Research and Technology Centre (FTZ) at Büsum, an institute of Kiel University. Multiple-day ship-based surveys and aerial surveys were carried out in the area of the *alpha ventus* windfarm and in a reference area to determine changes in seabird abundance and distribution patterns (◻ Fig. 5.3). Data from all EIA studies in and around *alpha ventus* was also evaluated. Flight heights were measured with a rangefinder, and detailed observations of seabird behaviour were recorded during boat counts. Digital aerial surveys using HiDef technology were conducted to test new survey techniques for German offshore EIA monitoring.

☐ Fig. 5.3 The razorbill (*Alca torda*) was one of the key species monitored at the *alpha ventus* offshore test site during the research on resting birds (photo: Mathias Putze).

Chapter 12: *Research on migratory birds*
One of the reasons for refusing approval for an offshore windfarm project is if the windfarm poses a threat to bird migration (see Section 5 (6) (2) of the German Marine Facilities Ordinance). In light of this, bird migration was investigated in four research projects carried out by Avitec Research GbR and the Institute of Applied Ecology GmbH (IfAÖ). Bird migration was investigated at the windfarm using video and heat imaging, as well as various radar systems to monitor birds encountering the rotorswept zone and to capture evasive bird movements. Various remote sensing techniques were installed on one of the wind turbines, the transformer station and the nearby FINO1 research platform. The aim was to determine to what extent the presence of the twelve *alpha ventus* wind turbines influences bird migration in the German Bight during spring and autumn. The fact that most migratory birds are fairly small in size and that two-thirds of them prefer to fly by night posed a special challenge for the investigations.

Chapter 13 and 14: *Research on marine mammals*
Research on marine mammals primarily focused on underwater noise. Investigations into the impacts of pile driving and operational noise on harbour porpoises were carried out by University of Veterinary Medicine Hannover. The presence of harbour porpoises was surveyed by means of aerial and boat counts in the pre-construction phase, during the construction period and in the first years of the operational phase. These surveys covered a large sea area. To account for possible changes in habitat use, a comprehensive passive acoustic monitoring programme was carried out. The collected data allows conclusions to be drawn as to the spatial and temporal scale of displacement caused by the installation of a windfarm.

A habitat modelling project was conducted by DHI / DHI-WASY GmbH to evaluate data on distribution and abundance of harbour porpoises gathered during EIA studies in accordance with StUK3. The EIA data and the StUKplus research data was then combined for long-term analysis and modelling of harbour porpoise distribution in the German Bight – the *alpha ventus* test site area and adjacent waters.

Chapter 15: *Research on underwater noise*
Construction and operation of offshore windfarms produces underwater noise that is potentially harmful to marine fauna. Most offshore wind turbines are installed by means of impact pile driving, which causes strong, impulsive noise. The foundations at the *alpha ventus* test site were driven up to 30 m into the seabed. Using hydrophones positioned in the water column, the underwater noise was measured before and during the pile driving work as well as during the operation of the wind turbines. The measurements were carried out by the Institute for Technical and Applied Physics GmbH (itap) during the construction of *alpha ventus* in 2009 and during the second year of operation in 2011. Most measurements were made with autonomous recording systems deployed 800 to 2,400 m from the *alpha ventus* test site. Two hydrophones were positioned in the 'Borkum Reef Ground' conservation area, at a distance of about 16 km from *alpha ventus*. As part of the project conducted by Müller-BBM GmbH, guidelines for underwater sound monitoring were developed. These 'measuring instructions' now form the standard for conducting underwater noise measurements during windfarm construction and operation.

◘ **Fig. 5.4** Colonisation of the foundation structures of offshore wind turbines was thoroughly investigated by scientific divers in the benthos research project (photo: Roland Krone).

5.3 Learning from the past to optimise future monitoring

At the beginning of the *alpha ventus* research project in 2008, the Standard for Environmental Impact Assessment currently in use was the second updated version (StUK3, BSH 2007). This already included outline monitoring designs for the construction and operation of a windfarm. However, the standard had never yet been applied in a real construction process. The monitoring concepts were so far proven for baseline studies, i. e. for monitoring studies before windfarm construction, and provided the legal assessment basis for windfarm licensing procedures. Applying StUK to a windfarm in construction and operation made it necessary to evaluate the study methods. Strategies for operational monitoring were absent and had to be developed.

One of the major goals of the StUKplus project was therefore to evaluate the StUK3 framework and to provide new information and insights for the next update of the standard, StUK4 (BSH 2013), which was released in October 2013. Based on the lessons learned from StUK3 monitoring and experience from the StUKplus research programme, new survey techniques and monitoring technologies were implemented and collected in a newly standardised form. For instance, the radar technologies for detecting bird migration newly applied during the StUKplus project are now part of the StUK4 monitoring design and aerial digital surveys for monitoring seabirds in windfarm areas are the only survey technique permitted for EIAs in German windfarm projects. Based on experience gained regarding the potential environmental impacts of offshore windfarms, the focus of the investigation scope was shifted from the all-round monitoring of earlier versions of StUK to more targeted monitoring in the current StUK4 – without losing the overall high quality standard of the StUK monitoring approach.

Literature

BSH (2013). Standard Investigation of the Impacts of Offshore Wind Turbines on the Marine Environment (StUK4). Bundesamt für Seeschifffahrt und Hydrographie, Hamburg and Rostock, 86 p.

BSH (2007). Standard Investigation of the Impacts of Offshore Wind Turbines on the Marine Environment (StUK3). Bundesamt für Seeschifffahrt und Hydrographie, Hamburg and Rostock, 58 p.

Conservation features: Sensitive species and habitats in the German Exclusive Economic Zone

Jochen Krause

Federal Maritime and Hydrographic Agency,
Federal Ministry for the Environment, Nature Conservation and Nuclear Safety (Eds.)
Ecological Research at the Offshore Windfarm alpha ventus,
DOI 10.1007/978-3-658-02462-8_6, © Springer Fachmedien Wiesbaden 2014

6.1   Introduction

The use of wind strongly and often constantly blowing over our seas and oceans has a long tradition. The modern use of wind as a renewable energy source in the form of offshore windfarms has only just started. In German seas, most of the windfarms – whether operational, under construction, approved or applied for – are located in the southern North Sea (see Chap. 1). From a biological point of view this part of the North Sea is one of the most productive parts of the Atlantic Ocean and historically it is known as a sea with an overwhelming number of species (Roberts 2007). The development of offshore windfarms is planned in a marine water body whose ecological condition has already been significantly altered by existing anthropogenic impacts, as a result of which conflicts with the diverse flora and fauna of the North Sea have increased continuously. Obviously, the North Sea gives the impression that it could provide space for a number of anthropogenic installations. However, approvals for such installations have to take into account the long history of human activities and their existing impact on species and habitats in the water column and on the sea floor.

6.2   Status of species and habitats of the North Sea

The current status of the German part of the southern North Sea is described in the national report on ecological status in accordance with Article 8 of the Marine Strategy Framework Directive (MSFD). Published in 2012, the report spans the preceding six years and also integrates assessments from a period before any windfarms were installed. The MSFD requires all Member States of the European Union to maintain and where necessary restore 'good environmental status' in the EU's marine waters by 2020. The German inventory for the 2012 report on the current status comprises a description of physical, chemical and biological status of the marine waters, i. e. the German parts of the North Sea and Baltic Sea, along with habitat types and hydromorphology. The initial assessment was limited due to inadequate data and a lack of assessment tools. Even so, several characteristic marine groups – phytoplankton, macrophytes, macrozoobenthos, fish, marine mammals, and sea birds – are found not to possess good environmental status (SRU 2012).

These results are not surprising: The OSPAR Quality Status report from 2010 for the North-East Atlantic including the North sea already stated that besides successes such as decreased pollution from oil and gas production, nutrient inputs have continued to increase, fishing and additional human pressures have large impacts on the marine environment and the decline in biodiversity is a long way from being halted. The OSPAR convention covers most of the North-East Atlantic and its adjacent seas. It is firmly rooted in global obligations and commitments such as the 1992 Convention on Biological Diversity (OSPAR 2010).

The OSPAR convention goes back to the 1970 s when two previous conventions, the Oslo Convention from 1972 and the Paris Convention from 1974, had the objective of protecting the North-East Atlantic from contaminants and nutrient discharges. Since 1998 the new OSPAR convention has included the protection of habitats and species by way of its Annex V. As an important step, the OSPAR Commission adopted in 2008 a list of threatened and/or declining species and habitats (OSPAR 2008) and is developing measures for the protection of those species and habitats. The two OSPAR lists include coastal habitats like zostera beds, offshore habitats such as sea pens and burrowing megafauna communities and species like the icelandic cyprine.

In addition to the OSPAR species and habitats, the EU Habitats Directive (Council Directive 92/43/ EEC, 1992), although mainly terrestrial in focus, aims to protect a number of marine species and habitats, which are listed in Annex I (natural habitats, i. e. coastal habitats such as mudflats, coastal lagoons, etc., and offshore habitats like sandbanks slightly covered by sea water all the time, reefs, etc.), Annex II (species, i. e. harbour porpoise, grey seal and andromous migratory fishes like twaite shad and sea lamprey) and Annex IV, which names species in need of strict protection, i. e. the harbour porpoise (◼ Fig. 6.1).

The EU Birds Directive (Council Directive 79/409/EEC, 1979) requires member states to undertake conservation measures for the habitats of

◻ **Fig. 6.1** Habitats and species protected by the EU Habitats Directive – (a) sublitoral sandbank with green mandarinfish (*Callionymus lyra*) in the North Sea, (b) reef with plumose anemones (*Metridium senile*) at the Sylt Outer Reef, (c) harbour porpoise (*Phocoena phocoena*), (d) grey seal (*Halichoerus grypus*); (photo: (a, b) Peter Hübner, Jochen Krause / BfN; (c, d) Katrin Wollny-Goerke / meeresmedien).

birds in danger of extinction listed in Annex I of the Directive and/or of regularly occurring migrating sea birds, amongst other things. The waters of the German North Sea are regularly used as a feeding, moulting or resting habitat by divers like the red-throated diver, the black-throated diver, well-known sea birds such as the northern fulmar, gannet, terns, including the arctic tern and the common tern, as well as the common guillemot, and gulls.

Last but not least, in 2010 Germany adopted a new Federal Nature Conservation Act (BNatSchG, 2009), section 30 of which designates five marine habitats as being strictly protected. These are 'sublittoral sandbanks', 'reefs', 'macrophyte meadows and beds', 'sea pen and burrowing megafauna communities', and 'species-rich shell, coarse sand and gravel bottoms'. These habitats basically reflect existing lists from the regional seas convention or the Habitats Directive. In Germany they were already classified in the national red lists of habitats (Riecken et al. 2006) as threatened and/or declining. Under section 30 (2) of the Federal Nature Conservation Act, activities that destroy or have a significant impact on such habitats are generally prohibited.

All these habitats and species are thoroughly described – including information on threats and sensitivities – in two books published by the Federal Agency for Nature Conservation (BfN) (Mendel et al. 2008, Narberhaus et al. 2012).

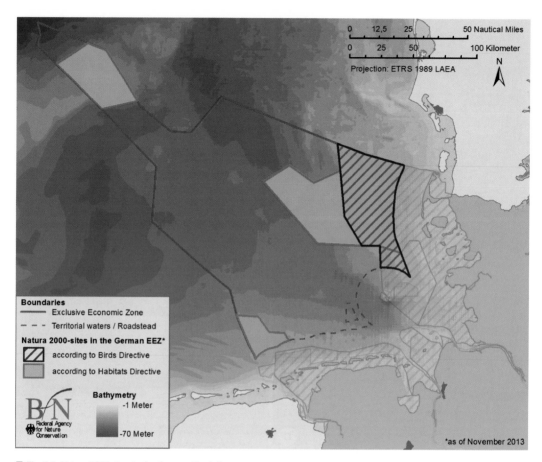

□ **Fig. 6.2** Natura 2000-sites in the German North Sea.

6.3   Impact of windfarms on species and habitats

The installation of offshore windfarms is not per se an additional threat for endangered habitats and species in the North Sea. Speaking in general terms, severity and duration of impacts depend on the relative amount of damage, disturbance or destruction caused by a single installation, as well as that caused by all installations cumulatively in respect of the total local population of a species or the total size of a habitat in the German North Sea. Garthe & Hüppop (2004) presented a method to develop a 'windfarm sensitivity index' and showed for the first time that bird species differ greatly in sensitivity for this specific threat. Black-throated divers and red-throated divers ranked highest (most sensitive), followed by velvet scoters, sandwich terns and great cormorants.

The smallest values were recorded for black-legged kittiwakes, black-headed gulls and northern fulmars. The locations of planned and existing windfarms therefore have to be analysed thoroughly in relation to the occurrence and distribution of these species and habitats. To analyse the distribution of habitats, the Federal Agency for Nature Conservation (BfN) together with the Federal Maritime and Hydrographic Agency (BSH) conducted a research project to map the sediments and the biotopes of the German North Sea. From 2002, distribution and abundance of sea birds and marine mammals were also sampled for the German North Sea (► www. habitatmare.de; Wollny-Goerke & Eskildsen 2008) This work means that windfarm licence approval proceedings can now include evidence-based analysis of the specific impacts of windfarms on species and habitats of concern.

6.4 The role of Marine Protected Areas

One important pillar in efforts to halt the loss of the marine biodiversity and give declining and threatened species and habitats time and space to recover consists of Marine Protected Areas (MPAs). Halpern (2003) showed that marine reserves worldwide and for all climate zones, regardless of size, and with few exceptions, lead to increases in density, biomass, individual size, and diversity in all functional groups, showing that MPAs are a significant tool for safeguarding threatened marine biodiversity. In recent decades, Germany has developed an impressive network of protected areas in the coastal waters and the Exclusive Economic Zone (EEZ) of the North Sea (Krause et al. 2011, von Nordheim et al. 2006). 43 % of the surface of the German North Sea is designated as Natura 2000 sites under the EU Habitats Directive (◘ Fig. 6.2). Most sites are additionally designated as OSPAR MPAs (OSPAR 2013). In the German North Sea, many, though not all, of the declining and threatened species mentioned above have an abundance peak within MPAs. Additionally, the mentioned habitats occur regularly within MPAs and most, though not all, have their main distribution area within MPAs. This does not automatically mean, however, that they are protected within the MPAs in the German EEZ. This is because designating legislation and management plans still have to be adopted or implemented for some MPAs. In 2009, however, a spatial plan for the German North Sea EEZ entered into force, prohibiting in particular the installation of windfarms within Natura 2000 sites (BSH 2009). Additionally, under the German Renewable Energy Sources Act (EEG 2012), there are no subsidies for windfarms approved in marine Natura 2000 sites after 2004. Due to the specific legal situation in the German North Sea, therefore, only a single windfarm has so far been granted a concession within an MPA, and no windfarms have been built on protected sea floor habitats or within the habitat of a protected species.

6.5 Conclusion

In conclusion, windfarms can have a negative impact primarily on threatened and declining species and habitats that have their distribution area outside the MPAs. Of particular concern are species and habitats that prefer fine sediments. Of only marginal concern are most coarse sand habitats and reefs as these are mainly found inside MPAs. An additional problem comprises barrier effects on migration routes, between stepping stone habitats, and impacts such as sediment plumes or underwater noise from pile driving, caused by windfarm construction outside of MPAs but affecting protected habitats and species within them.

In future, windfarm licensing proceedings will also have to include the currently incompletely implemented MSFD objective of attaining good ecological status. This means that in addition to the analysis of threatened and declining species, it will also be necessary to assess the effects on representative predominant habitats and functional groups of species. So far there are no such lists with final approval for Germany or the OSPAR region.

Literature

BNatSchG (2009). Bundesnaturschutzgesetz vom 22.07.2009. BGBl. 2009 I: 2542.

BSH (2009). Umweltbericht zum Raumordnungsplan für die deutsche ausschließliche Wirtschaftszone (AWZ) in der Nordsee. Bundesamt für Seeschifffahrt und Hydrographie (BSH), Hamburg and Rostock, 537 p.

Council Directive 92/43/EEC of 21 May 1992 on the conservation of natural habitats and of wild fauna and flora. (Habitats Directive).

Council Directive 79/409/EEC of 2 April 1979 on the conservation of wild birds. (Birds Directive).

EEG (2012). Gesetz für den Vorrang Erneuerbarer Energien (Erneuerbare-Energien-Gesetz – EEG) vom 25. Oktober 2008 (BGBl. I S. 2074), durch Artikel 1 des Gesetzes vom 17. August 2012 (BGBl. I S. 1754) geändert.

Garthe St, Hüppop O (2004). Scaling possible adverse effects of marine windfarms on seabirds: developing and applying a vulnerability index. Journal of Applied Ecology.

Halpern B (2003). The impact of marine reserves: Do reserves work and does reserve size matter? Ecological Applications, 13(1) Supplement, 2003: 117–137.

Mendel B, Sonntag N, Wahl J, Schwemmer P, Dries H, Guse N, Müller S, Garthe St (2008). Profiles of seabirds and waterbirds of the German North and Baltic Seas. 427 p.

Krause J C, Wollny-Goerke K, Boller F, Hauswirth M, Heinicke K, Herrmann Ch, Körber P, Narberhaus I, Richter-Kemmermann A (2011). Die deutschen Meeresnaturschutzgebiete in Nord- und Ostsee. Natur und Landschaft. 86 Jg. (9/10): 397–409.

Narberhaus I, Krause J, Bernitt U (2012). Threatened Biodiversity in the German North and Baltic Seas. Natursch. Biol. Vielf. 117. 628 p.

von Nordheim H, Boedeker D, Krause JC (2006). Progress in Marine Conservation in Europe. Springer. 260p.

OSPAR (2008). Case Reports for the OSPAR List of Threatened and/or Declining Species and Habitats. OSPAR Commission. 261 p.

OSPAR (2010). The OSPAR system of Ecological. Quality Objectives for the North Sea, a contribution to OSPAR's Quality Status Report 2010. OSPAR Commission.

OSPAR (2013). 2012 Status Report on the OSPAR Network of Marine Protected Areas. OSPAR Commission, Biodiversity Series. 64 p.

Riecken U, Finck P, Raths U, Schröder E, Ssymank A (2006). Rote Liste der gefährdeten Biotoptypen Deutschlands. Zweite fortgeschriebene Fassung 2006. Natursch. Biol. Vielf. 34. 318 p.

Roberts C (2007). The Unnatural History of the Sea. Island Press, Washington D.C.. 465 p.

SRU (2012). Umweltgutachten 2012. Verantwortung in einer begrenzten Welt. Rat von Sachverständigen für Umweltfragen. 422 p.

Wollny-Goerke K & Eskildsen K (Eds.) (2008). Marine mammals and seabirds in front of offshore wind energy. MINOS – marine warm blooded animals in North and Baltic Seas. Wiesbaden, 2008. 171 p.

Challenges, results and perspectives: An interview with Christian Dahlke

Federal Maritime and Hydrographic Agency,
Federal Ministry for the Environment, Nature Conservation and Nuclear Safety (Eds.)
Ecological Research at the Offshore Windfarm alpha ventus,
DOI 10.1007/978-3-658-02462-8_7, © Springer Fachmedien Wiesbaden 2014

1. Christian Dahlke, you were involved with *alpha ventus* from the outset, initially during the approval process. What did you see as the most pressing environmental issues back in 2001?

When the first applications to build offshore windfarms were submitted to the Federal Maritime and Hydrographic Agency (BSH) in 1999, it was clear that the marine environment would have to be a major focus in the investigations that lay ahead. At the time, I put forward a seemingly simple question: Why are ducks found in one place but not in another? Formulated like that, the question was new. It was soon clear that so far we only have a very vague idea of how the sea's various features and inhabitants interact – at least when it comes to projecting impacts of engineered structures on the marine environment. Many questions about the likely effects of windfarms on marine surroundings were not yet answerable with any certainty, but there were plenty of diffuse fears. Similar questions kept coming up in approval proceedings: How will resting birds react? Will migrating birds collide with the new obstacles at sea? What will be the effects of underwater pile driving noise and how can marine mammals be kept safe?

Then there was the question of uniform standards for offshore investigations – whether such standards were available in the first place. As many hours of discussion at application hearings showed, this was another area where new strategies and concepts were needed. Most of all, for many species there was a lack of monitoring methods. Driven by these needs, BSH worked with a large panel of experts to compile and publish the Standard for Environmental Impact Assessment (StUK).

2. The *alpha ventus* test site took on a pioneering role with regard to accompanying environmental research. What was achieved there that had not yet already been researched elsewhere? What was special?

While a lot of practical experience was available from elsewhere in Europe, concrete findings were hard to come by and even harder to apply to the German projects, which were further offshore and in deeper waters. Germany itself simply did not yet have any offshore windfarms or comparable offshore structures that could be used to verify projections made in EIAs. So the idea presented itself of making the approved Borkum West offshore windfarm, which was later renamed *alpha ventus*, into a reference project. The Federal Ministry for the Environment, Nature Conservation and Nuclear Safety (BMU) intended *alpha ventus* to set the ball rolling for German offshore development, and construction and operation of the test site were accompanied by extensive technical and environmental research. *alpha ventus* ultimately gave a first chance to study the environmental impacts in a real windfarm and so to gain a better understanding of processes in the marine environment.

3. What do you regard as the most important and successful outcomes from the StUKplus project?

First of all, there have not proven to be any severe harmful impacts on the marine environment. Now, it has to be said that *alpha ventus* is a small windfarm and there are still many open questions. That may well be the case, but it is something we knew beforehand. All the same, none of the fears from around the turn of the millennium have turned out to be grounded. This we know now, and it is quite something in itself.

But I also see another crucial point. If we want to know what impact not just one, but a large number of windfarms will have, we need a sound basis in the form of a clean monitoring standard developed in consensus. This is what StUKplus is about – about a sound StUK standard. No other country has that.

4. Where do you see a need for future research?

What I hope to see most is research on how we can apply innovative methods to even better mitigate the adverse impacts – most notably light and noise – of windfarm construction and operation so as to make this method of generating energy even more environment-friendly. And we need to give more attention than before to bird migration and to bats in the Baltic Sea.

Given the relatively small number of wind turbines in *alpha ventus*, it has not been possible to cover all conceivable impacts from the harvesting of offshore wind energy on an industrial scale. But a start has been made and the work should be continued, perhaps in further test sites. It would also be useful to see accompanying research and innovative technologies like gravity-based foundations.

Christian Dahlke
Former head of BSH department 'Management of the Sea',
Initiator of offshore windfarm approval procedures

Research in offshore areas

◘ **Fig. 7.1** Video and photo cameras were installed at FINO1 research platform for seabird registration (photo: Marine Measurement Net / BSH).

◘ **Fig. 7.2** Ornithologists at work: Counting seabirds at *alpha ventus* in good weather conditions (photo: Nicole Sonntag / FTZ).

◘ **Fig. 7.3** On foggy days at *alpha ventus*, counting of seabirds is difficult (photo: Nicole Sonntag / FTZ).

Fig. 7.5 Scientific divers took scrape samples at the foundation structures for biofouling analyses (photo: IfAÖ GmbH).

Fig. 7.4 Net caches with a pelagic trawl allowed identification of fish species composition and size distribution (photo: Kristin Blasche / BSH).

Fig. 7.6 The measuring chain at FINO1 platform was inspected on a regular basis (photo: Marine Measurement Net / BSH).

Fig. 7.7 Maintenance work in winter conditions (photo: Marine Measurement Net / BSH).

Main section

Oceanographic and geological research at *alpha ventus*: Instruments for predicting environmental conditions and interactions

Bettina Kühn & Anja Schneehorst

Federal Maritime and Hydrographic Agency,
Federal Ministry for the Environment, Nature Conservation and Nuclear Safety (Eds.)
Ecological Research at the Offshore Windfarm alpha ventus,
DOI 10.1007/978-3-658-02462-8_8, © Springer Fachmedien Wiesbaden 2014

8.1   Introduction

8.1.1   The RAVE Service Project

The last decade has seen growing demand for off-shore wind power. The weather conditions in the North Sea guarantee large wind energy yields. Yet strong winds lead to high and rough waves, increasing the costs and risks of windfarm construction and operation. Outside of the oil industry, there is little prior experience in building large structures in offshore conditions. What kinds of forces act on wind turbines? How often can vessels transfer people or material to and from windfarms? How can structural integrity be ensured in rough offshore conditions and do foundation structures cause changes in sediment structure? To answer these questions, the *alpha ventus* test site was dedicated as a playground for ongoing offshore research. Reliable information on the prevailing sea state – wave heights, wave directions and wave periods – is essential for planning offshore logistics and cost-intensive shipping and work operations. Such information is also essential for safety purposes. The transfer of people from a vessel to a wind turbine is one of the most critical situations for offshore engineers. At *alpha ventus*, the windfarm operators only allow access to wind turbines at wave heights less than 1.5 m (*alpha ventus* Health & Safety Executive, DOTI 2009). But even simple maintenance work is heavily affected by wind and waves (◘ Fig. 8.1).

The RAVE Service Project is coordinated by the Federal Maritime and Hydrographic Agency (BSH) and divided into several work packages covering a wide range of technical and scientific measurements. A central feature of the test site research was that measurements were taken directly from offshore structures. As many sensors as possible were mounted prior to erection of the turbines at sea to reduce the amount of difficult offshore installation work. Sensors under the water surface in particular had to be installed in advance onshore by the scientists. Technical measurements on the two different foundation types were conducted by the German Wind Energy Institute (DEWI) and GL Garrad Hassan Deutschland (GL). GL also conducted measurements at transmission level. BSH collected geological and oceanographic data to obtain reliable information about interactions between offshore facilities and the marine environment. The FINO1 research platform in the direct vicinity of *alpha ventus* was also part of BSH's oceanographic research programme. Meteorological, oceanographic and ecological data has been collected there since 2003, when the German government first decided to investigate the offshore environment and in particular the prevailing wind conditions (▶ Information box: *FINO1 research platform*). BSH also coordinated and supported all measurements for the RAVE research initiative in the test site area.

The focus of the geological investigation programme was on interactions between the piled foundations, the seabed and the oceanographic parameters (e.g. wave, tide and current), and sediment shifting in the windfarm. The depth and extent of scour can influence the stability of a structure. In the absence of in-situ measurements, a combination of fixed single-beam echosounders and full-coverage multibeam echosounding was applied to acquire a spatio-temporal dataset of scour dynamics at *alpha ventus*.

8.1.2   Environmental conditions in the North Sea

The prevailing weather conditions in the North Sea are ideal for the exploitation of renewable energy. Winds tend to be stronger and more sustained offshore than onshore. Atmospheric circulation over the German Bight is dominated by the Westerlies, hence the annual average wind direction is west or southwest (◘ Fig. 8.2).

The North Sea is situated on the European continental shelf. Water depths in the German Bight are around 40 m. The North Sea is open to the Atlantic Ocean and therefore dominated by a tidal signal entering from the northwest. The tidal wave circulates anticlockwise along the coasts. Its centre is the amphidromic point where no water elevation takes place and where there is no tidal range at all. Leaving the deep water and entering the continental shelf, the tidal currents increase and turbulent mixing occurs. Solar heating causes stratification of the water column, resulting in warm water in up-

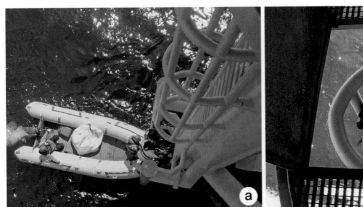

Fig. 8.1 (a) Transfer from zodiac to boat landing. (b) Engineer climbing down the boat landing (photo: Marine Measurement Network / BSH).

per layers and colder water below. These two drivers compete unequally as the tidal force is dominant. Stratification of the water column like that found in the Baltic Sea is therefore rare. Wind conditions modulate, increase or decrease the ocean currents driven by the tides.

The seabed of the North Sea was shaped during the last ice age more than 10,000 years ago. Today's seabed was part of the land masses and sea level was much lower than at present. After the melting of the glaciers, the world sea level increased, flooding the land masses and creating the North Sea. The sediments close to the *alpha ventus* test site are homogenous, their main components being fine sands that differ only in their content of fine material.

The geological model of the upper subsurface (**Fig. 8.3**) of the *alpha ventus* site is based on the results of a high-resolution sub-bottom profiler survey performed by the BSH vessel WEGA in 2008. The survey was carried out before construction at *alpha ventus* started. The seismo-stratigraphic model was set in relation to the sediment characteristics found from drillings taken during research for *alpha ventus*. It displays the geological features in the upper 5 m of the seabed, such as palaeochannels and trenches which are of relevance for understanding the scouring process.

8.2 Methods

8.2.1 Study design: Oceanography

A range of sensors and measurement devices were mounted on wind turbines and in the test site for detailed investigation of oceanographic conditions. Temperature sensors, directional wave riders, video cameras and current sensors captured data to analyse interactions between the windfarm and the ocean. The sensors had to be replaced every half a year, making it necessary to exploit spells of good weather. Different sensor types were used to compare different measurement techniques in order to obtain high-resolution, reliable and redundant datasets and to develop the best possible monitoring approach.

A directional wave rider collected sea state data and transmitted it immediately to a receiver so that the information would be directly available to windfarm operators. This information was invaluable as it meant that the decision on whether to send a vessel to a windfarm could be based on data updated every half hour. The buoy was attached to an anchor by an elastic rope. Passing waves lifted the buoy up and down, enabling the parameters wave height and length, period and direction to be registered and transmitted.

A radar gauge and video cameras were mounted on two wind turbines to capture selected high wave events. The cameras were triggered by the information collected by the wave rider. If the wave rider recorded waves over 2 m, a signal was sent to the cam-

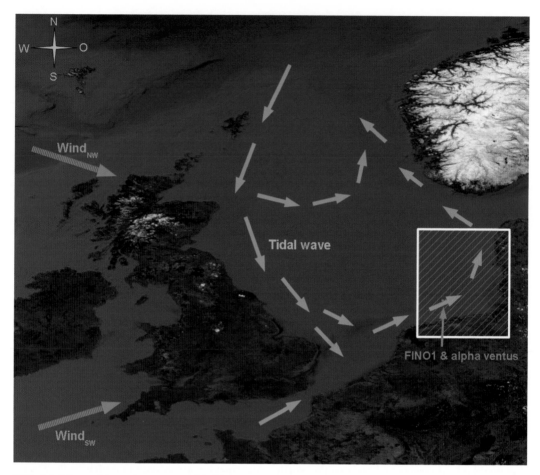

■ **Fig. 8.2** Scheme of the marine current model of the German Bight. Blue arrows indicate the tidal wave entering the North Sea and circulating anticlockwise. Green arrows indicate mean wind conditions.

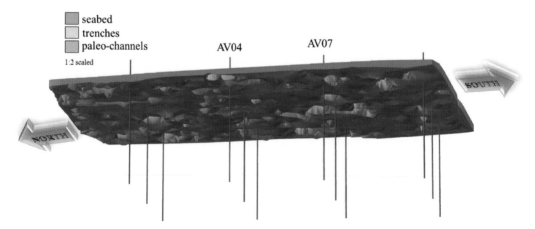

■ **Fig. 8.3** Geological model of the upper 5 metres of the seabed at the *alpha ventus* test site. Green indicates the seabed; yellow and dark orange indicate different palaeochannels; grey lines indicate the locations of wind turbines.

era, which then started recording. Independently, the camera captured waves of extreme heights using an event-driven mode. This worked by choosing an event window at a specific height (15 m above sea level); if there was any movement in the window, the camera started recording. The investigation of extreme wave events was important for safety questions and load calculations.

Acoustic Doppler current profilers (ADCP) collected current and sea state data and were placed in a frame on the seafloor close to the wind turbines. They were battery driven and had to be replaced periodically. An ADCP sends out a signal that is reflected back by small particles floating in different layers of the water column (such as plankton and floating sediment particles). A sensor calculates the current direction from the time it takes for the outgoing signal to be reflected back to the sensor. The sea state parameters are captured using a pressure sensor in the ADCP. When a wave passes, the pressure on the sensor rises or falls, allowing wave heights and periods to be calculated from the pressure differences.

◘ Figure 8.4 gives an overview of the sea state sensor positions in the test site and ◘ Fig. 8.5 shows a typical measurement setup on a wind turbine. Some sensors were battery driven, while others were cable-connected. The project started out with mainly cable-connected sensors but loss of cable capacity led to more battery-driven setups. This was a lesson that had to be learned – collecting data 75 km off the coast is not easy.

8.2.2 Study design: Geology

To monitor scour depths and processes, 33 single-beam echosounders were attached to a tripod (AV7) and a jacket foundation (AV4) before the turbines were erected at sea. At the beginning of the project, there were no single-beam echosounders for depths of 30 m on the market. It was therefore necessary to improvise using chandlery sensors. These were mounted in advance at the dockyards, so recording started very early, in 2009. Unfortunately, pile driving while erecting the wind turbines destroyed most

sensors on one of the wind turbines. Apart from the losses from pile driving, the sensors worked well in 30 m of water even though they were not designed for such depths.

From 2009, single-beam echosounders began recording data at high temporal resolution. A single-beam echosounder works by sending an acoustic signal vertically downwards to the sea floor and measuring the time that passes until its reflection comes back. The instrument than calculates the depth using the sound velocity based on temperature and salinity. A value was determined from five simultaneous measurements every 10 minutes, making it possible to record 120 values per day. The results were time-series for the several years since installation. They were corrected by a median filter and made it possible to determine both scour depth and scour dynamics. The echosounders at the tripod started recording in 2009 and those on the jacket foundations in 2010, the jacket foundations having been taken offshore later. As just mentioned, pile driving caused a lot of damage to the sensors, and nowadays only four sensors on the tripod foundations collect data. There are ten active sensors on the jacket foundations.

In addition to the single-beam echosounders, which only provide local measurements, we also acquired hydroacoustic measurements with multi-beam sonar (◘ Fig. 8.6). The aim of these measurements was to scan the seabed from a vessel moving at a distance of 40 m around the structure. The transmitter simultaneously sent out multiple beams in a fan-like formation transverse to the ship's motion. Coverage of the scanned area depended on the aperture angle and water depth. Scans were done in four transects until the bathymetric image of the seabed was complete. To obtain the vertical depths, the vessel's motion additionally had to be recorded. A motion sensor was therefore integrated into the measuring system. The sensor captured the vessel's drift, roll and pitch. A gyro compass was also incorporated to determine the fan orientation. Morphodynamic snapshots have been taken in this way at selected wind turbine locations every spring and autumn between 2009 and 2013 to record the scour situation before and after winter storms.

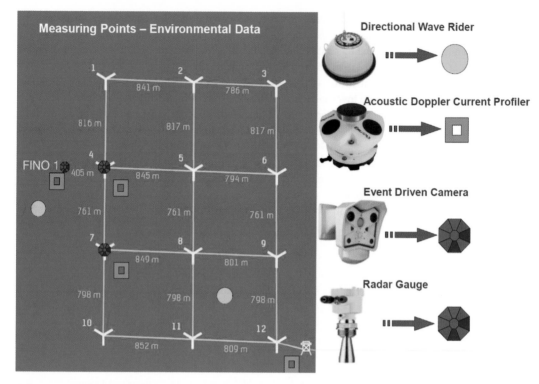

☐ **Fig. 8.4** Environmental measuring points in the *alpha ventus* test site, showing the four sensor types used to collect sea state data (during 2009–2013).

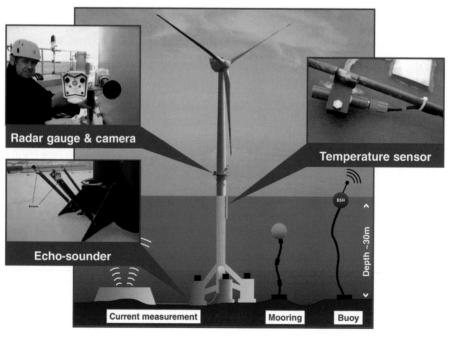

☐ **Fig. 8.5** Typical measurement setup for *alpha ventus*.

◼ **Fig. 8.6** Diagram of a multibeam echosounder measurement system.

Information box: How does scour develop?

The phenomenon of scour can occur in the surroundings of a structure located in a flowing medium such as the sea or a river. Changes in current patterns can result in increased erosion of the seabed: A flow pattern is disturbed by an obstacle (e.g. a rock or a pile foundation), passes around it and develops a horseshoe vortex (◼ Fig. 8.7b). This may cause depressions around the structure which are then called scour (◼ Fig. 8.7a).

Scour investigations are crucial in assessing the structural integrity of an offshore construction. Measurements of scour depths and dynamics are key parameters in periodic inspections by windfarm operators. The scope of measurements done during the RAVE Service Project went beyond the efforts of windfarm operators.

With regard to site investigations and construction of offshore wind turbines, initial experience showed that, as with environmental investigations, lack of experience and standardisation posed a challenge to licenses. To improve legal and investment certainty, BSH has published a standard on the subject. Compiled in cooperation with a group of experts, the standard lays down detailed minimum requirements for mandatory geological/geophysical and geotechnical site investigations at planned windfarm sites (Standard Ground Investigations for Offshore Windfarms, BSH 2008).

A further standard has been drawn up that specifies the requirements for offshore wind turbine

design and ensures that all installations and structural components are certified (Standard Design of Offshore Wind Turbines, BSH 2007).

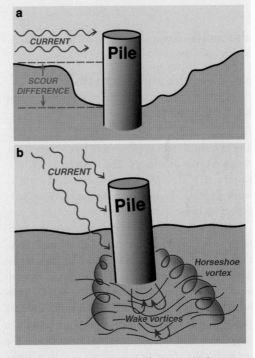

◼ **Fig. 8.7** Schematic diagram of scouring: (a) decrease in sediment, (b) correlation between current parameters and grain size distribution (after Hamill & Lucas 1999).

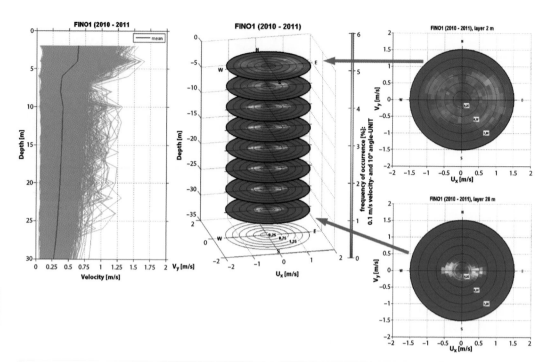

Fig. 8.8 Mean current velocities and directions, divided in 4 m layers, at FINO1 platform between 2010 and 2011 (graphs: Jens-Georg Fischer / BSH).

8.3   Results and discussion

8.3.1   Currents

The marine current profiles were taken by bottom-mounted ADCP sensors. The western and south-eastern ADCP were cable-connected (■ Fig. 8.5), while the ADCP close to the wind turbines were battery-driven.

The main currents at the *alpha ventus* wind-farm were heavily influenced in speed and direction by the tidal signal. ■ Figure 8.8 shows the mean current directions and velocities in the different water layers measured at the FINO1 platform between 2010 and 2011. The water column can be classified for illustration purposes into three main layers: In the upper layer (10 m below the sea surface), current flow was mainly driven by the wind and varied in direction and intensity. Down towards to the bottom, the water layers became heavily influenced in current direction and velocity by the tidal signal. Closest to the seafloor (5 m above the bottom), the current showed a strong east-west trend following the tidal rhythm. Close

to the bottom, friction played an important role and affected current velocity.

■ Figure 8.9 shows the bottom currents 5 m above the sea floor. The currents are split into their vector components with each subplot showing a different component. The two upper subplots show the horizontal current components. U represents the east (positive)-west (negative) current; v represents the north (positive)-south (negative) component. The third subplot shows the vertical velocity w, with positive values indicating movement to the surface and negative values indicating downward movement. The vertical velocities captured in *alpha ventus* were small but mainly negative. The tidal signal can be identified in the horizontal currents. A balanced current pattern can be identified in the upper subplot; the currents show the same intensity in the eastward and westward directions. The pattern in the subplot below is not as balanced, but the tidal rhythm can still be identified. The three current components varied strongly in intensity, but the identified current signal at *alpha ventus* matched with the overall current vector modelled for the North Sea, turning north-eastwards in the German Bight (■ Fig. 8.2).

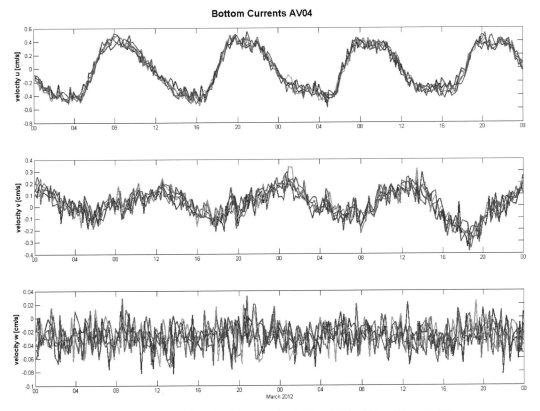

◼ **Fig. 8.9** ADCP time series for the bottom layer (5 m) over two days in March 2014 at 140 m distance to AV4.

8.3.2 Scouring effects

An impression of the scour depths and dynamics was gained by evaluating the time series from the single-beam echosounders. The diagrams (◼ Fig. 8.10a,b) show the maximum values from the measurement series. The individual time series for the different echosounders are partially comparable, but in principle each time series stands on its own. This is not surprising as scouring did not take place simultaneously at all locations.

In the case of AV7 (◼ Fig. 8.10a), one time series was included in the analysis for each pile. As can be seen, the NE and SE pile show similar results. The scour measurements for the west pile did not reach the same depths. The measurements for the central segment (C) correlated with the trends at the piles but not with the depths. At the end of 2012, the sensor broke down. All curves show a rapid increase in scour depth in the first half year. On average, scour increased by 40 cm per month

during the first half year. After that initial period, the scouring process slowed down. Since 2010 the average has been 5 cm per month. From 2009, local scours developed separately around each pile of the AV7 tripod structure. These evolved later on into one large scour beneath the whole tripod structure. From March 2010, the individual scours could hardly be distinguished.

The same analysis was done for the jacket foundation on AV4. The graph (◼ Fig. 8.10b) shows the time series for the north (N), east (E) and south (S) single-beam echosounders. Sonar at the western pile was unfortunately not working. Data were recorded at AV4 from September 2010. As with AV7, a significant scour developed in the first half year, with sediment removal at 30 cm per month. This process slowed down from 2011 with an average reduction of 2 cm per month. A look at the S curve in spring 2013 shows a sudden increase in scour depth of around 30 cm. It has not yet been possible to clarify why this happened. The scour hole is still subject

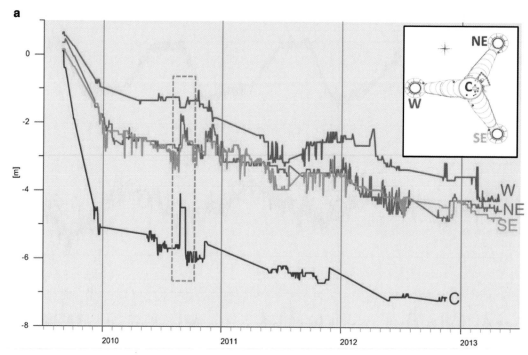

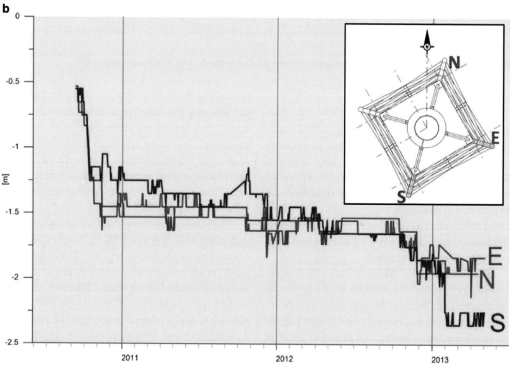

◻ **Fig. 8.10** Maximum values from the time series of single-beam echosounder measurements at (a) tripod structure AV7 (W: west pile; NE: northeast pile; SE: southeast pile; C: central segment) and (b) jacket structure AV4 (N: north pile; S: south pile; E: east pile) between 2009 and 2013.

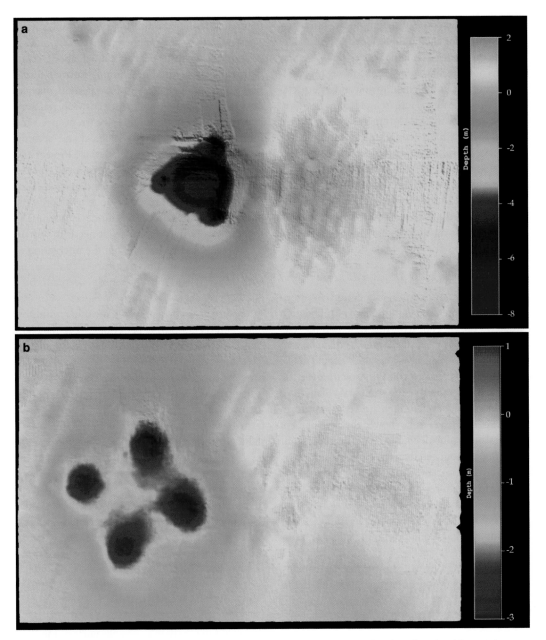

■ **Fig. 8.11** Diagram of the seabed topography and scour holing at AV7 (a) and AV4 (b) from boat-based multibeam echo-sounding surveys (March 2012).

to ongoing monitoring. All curves show short-term scour reductions with a maximum duration of two months (■ Fig. 8.10a, grey dotted box), but such reductions did not occur at all piles at the same time. It has not been possible to link these peaks to weather events, but it seems likely that there have been local sediment slips. The scour around the piles for the AV4 jacket structure (■ Fig. 8.11b) developed locally around each pile and has not so far evolved into a single scour.

The eroded sediment accumulated mainly to the east of both structures, although a slight accumulation of sediment to the west could also be observed.

The main outcome of the curves was to determine trends. The method developed here provides insights on scour development over time in a way that was not possible before. The scouring process is still in progress and its monitoring and interpretation is a task for future research projects.

The multibeam measurements made it possible to compile high-resolution elevation models. The survey results confirmed the data from the fixed single-beam echosounders (◨ Fig. 8.11a,b) and provided a broader view of the area of interest. Although the analysis was not continuous and only represented a snapshot of the scour situation, it contributed to a better understanding of the time series from the stationary single-beam echosounders. The method enabled the extent of the scour to be determined as well as the depth.

The results from both foundation locations demonstrated the different influence of foundation design on the effect of scouring. Based on the data from the geological subsurface model of *alpha ventus* (◨ Fig. 8.3), the deeper scour at AV7 could not be attributed to subsurface conditions within the test site area. It is notable that the scour at AV7 and AV4 is still increasing, even though at a very low rate. In the case of AV7, the results showed the development of a local and a global scour. At AV4, the local scour was more pronounced than the global scour.

Scour holing can be increased or reduced according to situation with regard to marine currents at the site in question. At *alpha ventus*, the tidal signal has a strong influence on current speed and direction, with a strong promoting effect on sediment transport. It can be concluded that different types of foundations will show different scouring behaviour depending on sediment, water depth and currents.

8.4   Perspectives

Many windfarms have been built in the North Sea in recent years and there are many more to come. The experience gained in the *alpha ventus* test site marked only the beginning, since many questions remain unanswered and new questions have arisen. The collected data has led to a better understanding of the marine environment. A need has been demonstrated for monitoring systems and online sea state data, as the weather changes very quickly offshore. The knowledge obtained about scour dynamic may be used in planning future foundations, and small changes in base designs may reduce material fatigue and hence cost. Correlations between sediment transport and current speed and direction were a focus of geological and oceanographic research in the RAVE Service Project and an insight was gained for the first time into the development of scours at offshore wind turbines. Further investigation needs to be done to identify the parameters responsible for scour development. As this process is highly complex, it needs to remain a focus of ongoing research, but the interaction of single-beam and multibeam sonars in combination with sea state and current sensors seems to be a good approach.

Future projects that investigate other wind turbines and their scouring behaviour are in the planning process. The sensors available on the market have improved and still need to be tested. To minimise the risk of being surprised by the weather and losing cost-intensive construction time, it is crucial to know about prevailing conditions in the windfarm area when planning offshore work. It is essential that the acquired experience and information be shared, not least because cost reduction remains a key factor for industry. Mistakes have to be made as part of the learning process but need to be kept to a minimum. Private-sector competitors should therefore work more closely together and communicate very openly on the problems they encounter in order to benefit from each other's knowledge.

8.5   Acknowledgements

We would like to thank our colleagues at the BSH Marine Measurement Network, the BSH Geology Sub-Division and the crews of the VWFS ATAIR and the VWFS WEGA. They all do a great job and without their capable help the project would not have collected so much data and been so successful.

Information box: FINO1 research platform

In 2002, the German government decided to erect dedicated research platforms in the North Sea and the Baltic Sea to improve knowledge of offshore conditions with a view to planned and approved offshore windfarm projects. Two research platforms were built in the North Sea: FINO1 next to *alpha ventus*, 45 km north of the island of Borkum (◘ Fig. 8.12), and FINO3, 80 km off the island of Sylt. The FINO2 research platform was erected in the southwestern part of the Baltic Sea. The FINO project is funded by the Federal Ministry for the Environment, Nature Conservation and Nuclear Safety (BMU). The FINO1 research platform started data acquisition in summer 2003 and a huge amount of environmental, meteorological, oceanographic and ecological data has been collected since. The platform is equipped with a wind met mast and physical oceanographic sensors under the water surface that capture temperature, conductivity, pressure and oxygen. Sea state data is collected with a sea state buoy and a radar gauge. Video cameras are additionally mounted on the platform to capture extreme wave events. The research focus is on optimising offshore logistics, developing safe shipping systems, detecting changes in ocean dynamics and developing meteorological and oceanographic model forecasts. Ecological research focuses on long-term monitoring of bird migration and artificial reef development on offshore structures.

BSH operates a large database of comprehensive meteorological and oceanographic measurements generated at the three research platforms. Use of the FINO data for research purposes is free of charge. For further information or data access, see ► www.bsh.de. For further information about the FINO project, see ► www.fino-offshore.de.

◘ **Fig. 8.12** FINO1 research platform (photo: Marine Measuring Network / BSH).

Literature

BSH (2008). Standard Ground Investigations for Offshore Windfarms. Bundesamt für Seeschifffahrt und Hydrographie, Hamburg and Rostock, 40 p.

BSH (2007). Standard Design of Offshore Wind Turbines. Bundesamt für Seeschifffahrt und Hydrographie, Hamburg and Rostock, 48 p.

DOTI (2009). Arbeitsschutz und Sicherheitskonzept zum Offshore Windpark *alpha ventus* der Betreibergesellschaft DOTI GmbH & CO KG, German Version, 72 p.

Hamill I & Lucas T (1999). Computational fluid dynamics modelling of tundishes and continuous casting moulds. Fluid Flow Phenomena in Metals Processing, pp. 279–286.

Rapid increase of benthic structural and functional diversity at the *alpha ventus* offshore test site

Lars Gutow, Katharina Teschke, Andreas Schmidt, Jennifer Dannheim, Roland Krone, Manuela Gusky

Federal Maritime and Hydrographic Agency,
Federal Ministry for the Environment, Nature Conservation and Nuclear Safety (Eds.)
Ecological Research at the Offshore Windfarm alpha ventus,
DOI 10.1007/978-3-658-02462-8_9, © Springer Fachmedien Wiesbaden 2014

9.1   Introduction

The benthos of the German North Sea – consisting of both invertebrates and fishes associated with the seafloor – is characterised by notable biodiversity. Hundreds of species from almost every taxonomic invertebrate group exist in complex interactions with their biotic and abiotic environment, thereby providing important ecosystem goods and services. Specifically, in the sandy sediment of the *alpha ventus* windfarm, about 200 benthos species form the typical *Tellina fabula* association named after one of its dominant bivalve species (Salzwedel et al. 1985). Many of these organisms are representatives of basal levels of the marine food web. Accordingly, they form an essential link between marine primary production by planktonic microalgae and consumers from higher trophic levels such as birds, mammals and commercially valuable fishes (Gili & Coma 1998).

Because of the eminent ecological importance of benthic organisms, national nature conservation authorities have emphasised the need for careful research into the potential effects of offshore windfarms on the marine benthos (Merck & von Nordheim 2000). The Standard for Environmental Impact Assessment (StUK), compiled by the Federal Maritime and Hydrographic Agency (BSH), prescribes extensive studies on the potential effects of offshore windfarms on the in- and epifauna of the seafloor as well as the fouling assemblage on the foundation structures of the turbines (see Chap. 5). This chapter presents the results from research projects in which selected aspects of the StUK3 investigation programme were temporally intensified. Additional sampling campaigns were added to close temporal gaps in the seasonal sampling programme. Special emphasis was given to the investigation of the mobile demersal megafauna on and around the underwater structures of the turbines. This component of the benthos was studied by scientific SCUBA diving – a method which is very challenging under the rough offshore conditions of the North Sea, but provides unique results that cannot be obtained with any other method (► Information box: *Scientific offshore diving*). In interpreting the data from our benthos survey we drew upon a unique extensive dataset on the distribution of benthic species in the German Bight. This allowed an evaluation of the specific development of the benthic community in the *alpha ventus* windfarm area against the ambient large-scale variability of the benthic ecosystem caused by natural environmental fluctuations in combination with anthropogenic stressors such as bottom fisheries. In sum, common methods were applied in the investigation of the marine benthic in- and epifauna, standardised national strategies for environmental impact assessment modified, and new additional methods were included to provide a comprehensive understanding of the effects of offshore windfarms on the marine benthos.

9.2   Methods

The study objects, the sampling designs and the survey methods were mainly based on the StUK3 procedures (BSH 2007; ◘ Fig. 9.1). Additional new methods were adopted to sample the mobile demersal megafauna, including large crabs and demersal fish that occur around the underwater structures of the wind turbines and on the adjacent seafloor.

9.2.1   Study design

One of the most widely used designs was applied to sample the benthic in- and epifauna: A modification of Green's (1979) Before-After-Control-Impact (BACI) design. This involves sampling the benthic fauna in an area that is planned to be affected by some disturbance and also in a single reference (control) area not affected by the development. Each area is sampled once prior to and several times after the potential disturbance.

Accordingly, the benthic fauna was surveyed within the *alpha ventus* area and also in a reference area outside the windfarm (◘ Fig. 9.2). The reference area was similar to the windfarm area in terms of abiotic parameters such as water depth, type of sediment and biotic features (e.g. species inventory of the benthic community). Thus, the reference area served as a control to identify possible effects of the construction and operation of the windfarm on the temporal development of the benthic fauna.

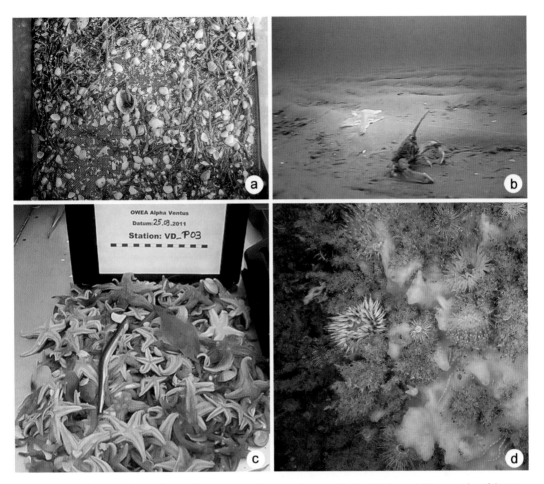

Fig. 9.1 Typical catches taken with the various survey methods used to study the benthic fauna. (a) Sieve residue of the van Veen bottom grab characterised by bivalves and tubeworms. (b) Video recording of sedimentary seafloor with the masked crab *Corystes cassivelaunus* and a starfish. (c) Typical beam trawl catch with numerous starfish *Asterias rubens*. (d) Characteristic fouling community on the foundation structure of research platform FINO1 which is located close to the *alpha ventus* windfarm (photo: (a-c) IfAÖ GmbH, (d) AWI).

After the baseline survey in spring 2008, which was conducted prior to construction of the offshore wind turbines in *alpha ventus*, the soft-bottom fauna was sampled biannually from 2009 until 2011, with sampling campaigns in spring and autumn. The assemblages of fouling organisms, such as mussels and sea anemones (**Fig. 9.1d**) that colonised the turbine foundations, were sampled for the first time after construction of the wind turbines in 2009. Repeated samplings followed in spring and autumn 2010 and 2011. The mobile demersal megafauna was sampled in spring and autumn 2011, i. e. two years after construction of the windfarm.

9.2.2 Data collection

Soft-bottom communities

The soft-bottom epifauna, i. e. often mobile invertebrate organisms that live on the seafloor, such as crabs, shrimps and starfish (**Fig. 9.1b**), was sampled with a beam trawl with an opening width of 200 × 60 cm and a mesh size of 1 cm (**Fig. 9.3**). Sampling was always done during daytime. For each epifauna sample, the beam trawl was towed on the ground for five minutes at a trawling speed of 1 to 3 kn. For each campaign, ten beam-trawl samples were taken within *alpha ventus* and in the reference area (**Fig. 9.2**), resulting in a total of 140 hauls be-

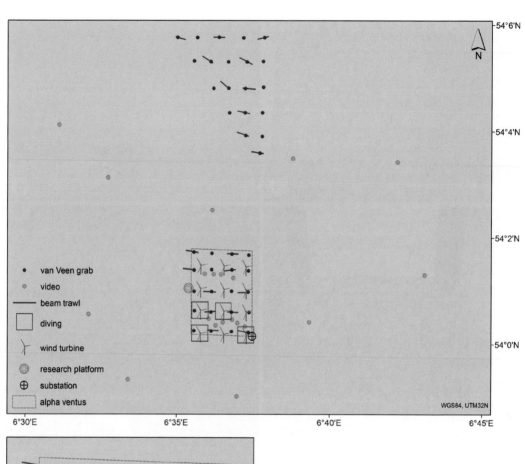

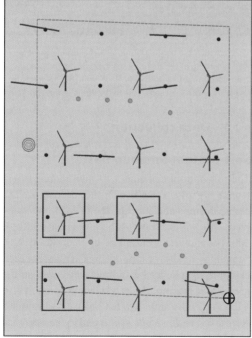

Fig. 9.2 Geographic positions of the stations within and outside (reference) the *alpha ventus* windfarm, where the benthic fauna was sampled from 2008 until 2011 by various survey methods.

Fig. 9.3 Schematic depiction of the survey methods applied to achieve a thorough quantitative assessment of the benthic fauna at *alpha ventus*: (1) Epifaunal inhabitants of the sediment were collected by bottom trawling. (2) Video recording was carried out to obtain an overview of the mobile demersal megafauna on the sedimentary seafloor. (3) The infauna was sampled by using the van Veen bottom grab. (4) Scientific diving is an excellent method for sampling the fauna associated with the artificial hard substratum of the wind turbine foundations.

tween 2008 and 2011. On board the research vessel, the catch was sorted, counted and weighed (wet weight) by taxa.

The soft-bottom infauna, i. e. the animals that live inside the sediment, such as clams, tubeworms and burrowing crabs (☐ Fig. 9.1a), was sampled

with a van Veen bottom grab with a sampling area of 0.1 m^2 and a weight of 75 kg (☐ Fig. 9.3). On each sampling campaign, 20 evenly distributed stations were sampled in the *alpha ventus* area and in the reference area (☐ Fig. 9.2). Two replicate grabs were taken at each station, resulting in a total of 560 samples between 2008 and 2011. After the van Veen grab had been lifted from the seafloor, it was flushed out immediately with seawater and the sediment was sieved through a 1 mm mesh. The retained animals were preserved with 4 % borax-buffered formalin in seawater. The preserved infaunal organisms were then identified, counted and weighed (wet weight) by taxa in the laboratory.

Hard-bottom associated fauna

The hard-bottom fauna on the turbine foundations was sampled by scientific divers (◘ Fig. 9.3). The scientific diving operation is described below, ▶ Information box: *Scientific offshore diving*.

During each sampling campaign, the fouling organisms were sampled on two wind turbine foundations (◘ Fig. 9.2). A total of 126 samples were taken between 2008 and 2011. On each turbine foundation, three replicate scrape samples were taken randomly at water depths of 1, 5 and 10 m, respectively. The fouling organisms were scraped off with a spatula from a 20 × 20 cm area and captured in a mesh bag (mesh size: 1 mm). The organisms were fixed with 4 % borax-buffered formalin. The preserved individuals were then identified and weighed (wet weight) in the laboratory.

The mobile demersal megafauna was surveyed visually on belt transects extending on the seafloor away from three to four turbine foundations (◘ Fig. 9.2). Additionally, the megafauna was recorded on the three-dimensional underwater structure itself. All individuals were identified *in situ* and counted by the diver. Each record was reported to a

co-worker on the surface via underwater telephone. The megafauna abundance from the surveyed area of the underwater structure was extrapolated to the entire subtidal area of the foundation. The megafauna abundance from the belt transect was extrapolated to the projection area of the foundation on the seafloor, plus the surrounding 15 m transect belt resulting in what is referred to as the 'footprint area' of the turbine foundation measuring 2,400 m². Adding up the abundances of the mobile demersal megafauna from both the footprint area and from the turbine foundation allowed an estimate to be made of the overall amount of megafaunal individuals that accumulate above the footprint area of a wind turbine foundation.

As a reference for mobile demersal megafauna at the turbine foundations, the megafauna was surveyed on the sedimentary seafloor inside and outside the windfarm area using a ship-based underwater video camera system (CMOS video TV resolution; 9 W high power LED light; ◘ Fig. 9.3). The camera was towed above the seafloor at a drift speed of 0.2–0.5 knots. In total, 20 video transects of 500 m length each were generated (◘ Fig. 9.2). The camera was equipped with parallel lasers which allowed for counting the animals on strips of defined width and length. Sections of the strips were randomly selected from each video to obtain transects of the same projection area as a dive transect, resulting in a total of 110 seafloor transects of 15 m² each. The comparability of dive and video transects was confirmed by Krone et al. (2013a).

Scientific offshore diving

Personnel demands

When diving for scientific purposes using self-contained underwater breathing apparatus (SCUBA), German researchers must follow the national hazard prevention regulation. This regulation entails safety requirements with regard to equipment and personnel. It requires that the diver takes an exam and qualifies in scientific diving, and prescribes that a scientific diving crew consists of at least an instructor, a diver and a safety diver.

Challenges offshore

Research at offshore windfarms in the North Sea involving scientific diving requires additional effort. Given the water depth, often poor underwater visibility, and the possibility that divers may become entangled in submerged structures, the diving operation should be conducted 'surface supplied' (umbilical with air and telephone; ◘ Fig. 9.4). Furthermore, windfarm operators have their own specific demands for diving operations, and the use of offshore occupational diving equipment is recommended. Divers must be educated and trained at regular intervals in surface-supported diving at larger depths and also in seamanship. In tidal seas such as the North Sea, diving is most secure during the short slack water periods. Thus, in most instances, only two dives are possible during daylight in the North Sea. A sea-worthy expedition ship and temporally flexible expedition planning are essential as the success of a dive mission depends on suitable sea conditions. The personnel and material efforts of scientific offshore diving therefore have only little in common with snorkelling or diving for recreational purposes in shallow coastal waters.

Opportunities

If the challenges of scientific diving are mastered and expeditions are run as scheduled, unique new observations from 'exotic' sites such as wind turbine foundations are possible. Unlike in studies of cultured organisms in aquaria, the behaviour of almost undisturbed animals can be studied in situ. Diving allows for non-destructive quantification of the fauna in habitats where the use of remotely operated devices is difficult. This enables us to create a more comprehensive picture of North Sea artificial reef ecology by capturing species that would otherwise remain hidden and by estimating their abundance and distribution. Moreover, in situ observation by means of diving allows for more than just numerical information on habitats und their inhabitants. The diver is the most multi-functional tool in the sea, and a diving marine researcher experiences personally how aquatic organisms might live in a three dimensional space with reduced gravity and elevated pressure.

◘ **Fig. 9.4** Surface supplied equipment used for scientific diving operations to survey the fauna associated with the turbine foundations of the *alpha ventus* windfarm.

9.2.3 Data analysis

To evaluate potential windfarm effects on the marine benthic fauna, the data were analysed using common statistical methods (see e.g. Legendre & Legendre 1998). Univariate analyses, such as repeated measurements analysis of variance (rm-ANOVA), were carried out by means of the STATISTICA v7.1 software package (Hill & Lewicki 2007).

9.3 Results and discussion

To describe the response of the marine benthos comprehensively, various components of the macro-zoobenthos were studied, which were expected to react to both the construction and operation of the wind turbines at *alpha ventus*: The in- and epifauna of the sedimentary seafloor, the fouling assemblage on the underwater construction of the turbines, and the mobile demersal megafauna, which aggregated at the artificial structures. The applied BACI design is a common tool for the study of human impacts on natural communities (Green 1979). Proper statistical analysis of the collected data allows for (1) identification of an overall difference between the windfarm area and the reference area with regard to a selected variable such as species richness, total abundance or biomass of a species community. The analysis also tests for (2) any overall temporal variation in each variable in the entire area of investigation without distinguishing between the windfarm and the reference area. Finally, the analysis looks for (3) any interaction between the two factors 'area' (the comparison between the windfarm and the reference area) and 'time' (detection of temporal variation). An interaction between the two factors occurs if the temporal development of the variable is different in the two areas and is, thus, indicative of altered environmental conditions in the impact area given that the two areas are well comparable, i. e. the areas exhibit similar environmental conditions except for the focal disturbance. The reliability of this procedure rises with the degree of spatial and temporal replication in the sampling. Hence, optimal sampling would include several sampling campaigns before and after the impact, as well as several reference (and impact) areas to representatively portray the temporal and spatial variability of the community parameters (Underwood 1991, 1993). The sampling campaign prior to construction of *alpha ventus* in spring 2008 revealed minor differences between the prospective windfarm area and the reference area with regard to the sedimentary conditions and the structure of the local benthic communities. Nonetheless, the two areas were considered well comparable because the sediments were generally of the same type (i. e. homogeneous fine sand with low organic content).

The BACI approach allowed the identification of differential temporal variations in benthic communities in the *alpha ventus* area and in the reference area. However, the procedure is not able to clearly distinguish between direct impacts of turbine foundations (e.g. import of biomass into the sediment) and other processes associated with operation of the windfarm such as recovery of the benthic community after the cessation of bottom

trawling in the windfarm area. The term 'windfarm effect' is therefore used here to refer to the gross effect arising from various interactive processes that occur within the windfarm area but not in the reference area.

9.3.1   Epifauna

Species richness of the epifauna on the sedimentary seafloor was generally higher in the reference area than in the *alpha ventus* area (▣ Fig. 9.5a). The temporal variation of overall species richness during the investigation period was remarkable in both areas. The average number of species per haul was lowest in spring 2009 (4.9 ± 0.8 and 5.5 ± 0.8 in the *alpha ventus* and the reference area, respectively) and highest in the following spring 2010 (8.8 ± 1.1 and 10.5 ± 1.4 in the *alpha ventus* and the reference area, respectively). However, the temporal fluctuations in the number of species were largely synchronous in both areas, indicating that neither construction of the turbines nor the subsequent two-year operation period affected the development of species richness in the windfarm area. The total abundance of the epifauna increased slightly in both areas between 2008 and 2011 (▣ Fig. 9.5b), with the starfish *Asterias rubens* being the dominant epifaunal species in both areas. The temporal variation of total abundance was different between the two areas indicating windfarm effects on the development of the epifaunal abundance. Especially by the end of the investigation period, from spring to autumn 2011, total epifaunal abundance in the two areas progressively diverged, indicating a persistent difference in the communities. In fact, a recent sampling in spring and autumn 2012 confirmed that the abundance of epifauna increased further in the reference area whereas in the windfarm area abundance remained relatively stable (A. Schmidt pers. obs., data not shown here). Future sampling campaigns will reveal whether the numbers of individuals will diverge further or whether this was just a transient development. Total biomass of the epifauna also fluctuated heavily throughout the study period (▣ Fig. 9.5c). As for total abundance, fluctuations differed in the two areas, again indicating windfarm effects on the benthic epifaunal community structure.

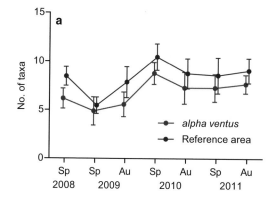

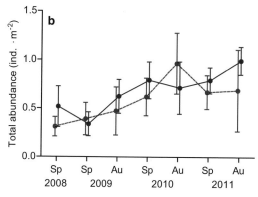

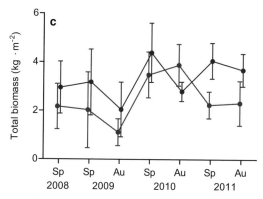

Fig. 9.5 Average (± standard deviation) (a) number of taxa, (b) total abundance and (c) total biomass of the benthic epifauna of the sedimentary seafloor inside *alpha ventus* and in a reference area outside the windfarm. In 2008, a single sampling campaign was conducted in spring (Sp). From 2009 to 2011, samples were taken semi-annually in spring (Sp) and autumn (Au).

9.3.2 Infauna

Similarly to the epifauna, the results for the infauna indicated windfarm effects on the benthic community structure. For all selected community descriptors (species richness, total abundance and total biomass), temporal development differed in the windfarm area and in the reference area. Species richness was higher in the reference area when the number of species was integrated over the entire investigation period (◘ Fig. 9.6a). Contrastingly, total abundance was initially higher in the *alpha ventus* area. However, by the end of the study period in 2011, this relationship reversed so that total abundance was finally higher in the reference area (◘ Fig. 9.6b). In both areas, the total abundance of the infauna was dominated by the highly abundant bristle worm *Spiophanes bombyx* so that the varying abundances were primarily the result of differential population dynamics in this single species. It is difficult to assign these differential dynamics solely to effects of the wind turbines in the *alpha ventus* area because the observed fluctuations were well within the range of natural population fluctuations for this species in the German Bight. Similarly, the temporal dynamics of the total infaunal biomass were dominated by the biomass of a single species, the sea urchin *Echinocardium cordatum*. Interestingly, the biomass of the urchin and, thus, the total infaunal biomass were substantially higher in the reference area than in the windfarm area only during the spring campaigns, whereas during the autumn campaigns the biomass was similar in both areas (◘ Fig. 9.6c). This specific pattern might be the result of the specific mating behaviour of this urchin. These sea urchins are known to aggregate in spring and summer for mating (Buchanan 1966). Along with the preference of this species for sediments with elevated organic content (Wieking & Kröncke 2003), this behaviour might result in the observed spring aggregation of the urchins in the reference area where the organic content of the sediment was slightly higher than in the windfarm area.

In accordance with the principles of environmental impact assessment using BACI, the differential temporal variations of various community descriptors in the windfarm and in the reference area indicated differential environmental conditions in

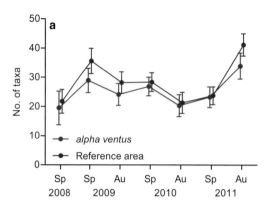

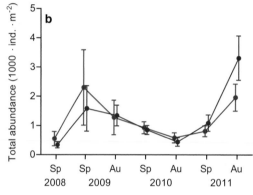

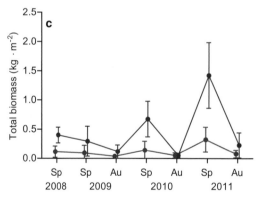

☐ **Fig. 9.6** Average (± standard deviation) (a) number of taxa, (b) total abundance and (c) total biomass of the benthic infauna of the sedimentary seafloor inside *alpha ventus* and in a reference area outside the windfarm. In 2008, a single sampling campaign was conducted in spring (Sp). From 2009 to 2011, samples were taken semi-annually in spring (Sp) and autumn (Au).

of the two areas as well as the inhabitant in- and epifaunal communities were similar, it cannot be entirely precluded that the differential community dynamics were the result of weak yet fundamental differences between the areas. In practise, it is impossible to find two entirely equal areas that would allow for straightforward identification of environmental impacts. Therefore, an optimised sampling would consider various suitable reference areas (Underwood 1993). The communities within these several areas are likely to display the entire range of ambient variation to which the variations observed in an impact area can be compared. An environmental impact could then be detected reliably when the temporal variation in the impact area exceeds the full range of ambient variation displayed by the communities of all reference areas. Additionally, an optimised sampling would include several sampling campaigns before the onset of an impact in order to obtain an estimate of the fundamental differences between impact and reference areas without the influence of the expected disturbance (Underwood 1992). To estimate the entire range of ambient fluctuations, a comprehensive data set was used on the abundance of the benthic fauna in the German Bight. As part of a project on the use of the benthos in marine spatial planning and permit procedures for offshore windfarms (funded by the BSH and the Federal Ministry of Transport, Building and Urban Development; grant no. 10016990), this dataset was used to estimate the variability of species richness and total biomass of the infauna on sediments which are comparable to the fine sand areas studied here (☐ Fig. 9.7). The variations were expressed by the coefficient of variation (c_v) which was calculated as $c_v = \sigma/\mu$ with σ being the standard deviation and μ the mean of the respective variable. The coefficients of variation for the benthic infauna in the *alpha ventus* area were 0.2 and 1.2 for species richness and total biomass, respectively. For the infauna in the reference area, the coefficients were 0.3 and 1.1 for species richness and total biomass, respectively. A comparison shows that the variations in both the windfarm and the reference area were well within the range of the ambient variations of the infauna on fine sand sediments in the German Bight (☐ Fig. 9.7). Hence, the infauna in the windfarm area did not show particularly strong

the two areas, which might have been the result of the construction and operation of the turbines in the *alpha ventus* area. However, the results have to be interpreted carefully. Although the sediments

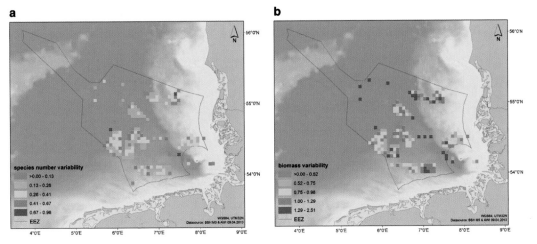

◻ Fig. 9.7 Variability (calculated as coefficient of variation; for details see text Sect. 9.3.2) of (a) species number and (b) total biomass of the benthic infauna on sublittoral fine sands in the German Bight.

variation that would indicate exceptional temporal development of the community due to the presence of the wind turbines.

In sum, the results on the benthic in- and epifauna of the sedimentary seafloor indicate windfarm effects on benthic communities or single species thereof. However, continued seasonal sampling will be necessary to confirm the potential effects. Studies from other European windfarms confirm the lack of short-term effects of offshore windfarms on the marine soft-bottom benthos (Lindeboom et al. 2011). However, in Belgium's Thorntonbank windfarm, structural changes of the benthic communities were evident six years after construction of the turbines and the results indicate a slow but persistent spatial expansion of the effects (Coates et al. 2012). It remains to be seen whether the effects will expand over the entire area of the windfarm or remain restricted to the vicinity of the turbine foundations.

9.3.3 Biofouling on the foundation structures

The investigations on biofouling provide the first continuous description of the temporal development of fouling biomass on the underwater structure of offshore wind turbines from the time of construction onwards. The results show that artificial hard substrata in the North Sea become colonised by a rich fouling community that can reach a substantial biomass. After construction of the turbine foundations in 2009, the average species richness of the fouling assemblage on the underwater structures increased steadily in water depths of 1 to 10 m, from about five taxa per sampled area (i. e., 0.12 m²) in autumn 2009 to about 15 taxa per sampled area by the end of the study period in autumn 2011 (◻ Fig. 9.8a). Dominant taxa (in terms of biomass) of the fouling assemblage were the blue mussel *Mytilus edulis*, the amphipod crustacean *Jassa herdmani* that builds extensive tubes on the surface of the substratum, anemones, and the starfish *Asterias rubens*.

The biomass of the fouling assemblage was not evenly distributed relative to the water depth. At 5 and 10 m depths, the average biomass hardly exceeded 1 kg m⁻² throughout the entire study period. In contrast, at the 1 m depth level the average biomass of the fouling assemblage increased rapidly from about 1 kg m⁻² in spring 2010 to about 25 kg m⁻² in autumn 2011 (◻ Fig. 9.8b).

The massive increase in fouling biomass at 1 m depth was primarily due to colonisation of the turbine foundations by the blue mussel. A similar development of mussel biomass was observed on the underwater structure of the FINO1 research platform, which is located next to *alpha ventus* (Joschko et al. 2008). Two years after the construction of the platform, the mussel population had risen to about

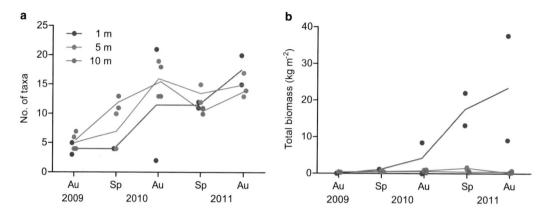

◻ Fig. 9.8 (a) Number of taxa and (b) total biomass of the community of fouling organisms on the turbine foundations at *alpha ventus*. Samples were taken in 1, 5 and 10 m water depth. Solid lines connect the seasonal means of two replicate samples from the same depth level. Samples were taken semi-annually in spring (Sp) and autumn (Au).

35,000 individuals m^{-2} and a biomass of 40 kg m^{-2}. Accordingly, the mussel population on the turbine foundations in the *alpha ventus* windfarm area has likely continued to increase since the final sampling in autumn 2011. Contrastingly, at 5 and 10 m depths no substantial biomass increase is expected for the future. In these water depths, the fouling biomass on the FINO1 platform appeared to be largely stabilised at an average of about 1 to 4 kg m^{-2} four years after construction (Krone et al. 2013b).

With regard to species composition, biomass and functionality, the fouling assemblage on the turbine foundations differed substantially from the common benthic communities of the surrounding sedimentary seafloor. For example, many species of the fouling assemblage are suspension feeders which exploit other food resources compared with the numerous benthic deposit feeders in the sediments. In comparison with the findings from previous studies on the fauna on wrecks in the North Sea (e.g. Zintzen et al. 2006), it becomes evident that offshore wind turbines differ from wrecks. Turbine foundations extend through the entire water column, providing settling space in shallow waters for mussels which are largely absent from wrecks on the seafloor (Krone et al. 2013b). Given their enormous biomass volumes, mussel aggregations on the underwater structures of offshore facilities are expected to have substantial impact on the fluxes of energy and matter, at least on a local scale (Joschko et al. 2008).

9.3.4 Mobile demersal megafauna

The large turbine foundations and, above all, the associated fouling organisms, attract huge numbers of mobile demersal megafauna. This faunal group consists mainly of demersal fishes and large decapod crustaceans known to aggregate at solid structures in the marine environment, probably for feeding, mating and/or shelter (Page et al. 1999, Sadovy & Domeier 2005). The fouling community on the turbine foundations provides a valuable food resource for large predatory species. Additionally, dead fouling organisms fall off their substratum and end up on the seafloor where they are foraged upon by benthic scavengers. During the scientific diving operations, 16 species of mobile demersal megafauna associated with the turbine foundations were observed versus 18 species on the nearby sedimentary seafloor. However, the much higher sampling effort on the sedimentary seafloor increased the probability of encountering additional species. Accordingly, we expect the species richness of the mobile demersal megafauna on the turbine foundations to be at least as high as on surrounding sediments. Much more striking than the comparison of the species richness is the comparison of the megafaunal abundance between the sedimentary seafloor and the artificial structures. For some species that aggregated on the turbine foundations, such as the edible crab, *Cancer pagurus*, and the common hermit crab, *Pagurus bernhardus*, abundances were

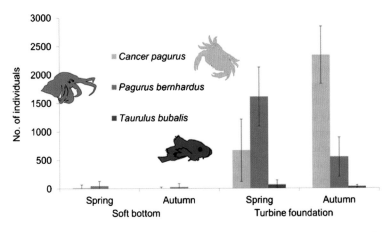

Fig. 9.9 Average (± standard deviation) abundance of selected species of the mobile demersal megafauna on the footprint area (2,400 m²) of the tripod foundations at *alpha ventus* and on soft-bottom areas of the same size. Samples were taken in spring and autumn 2011.

several orders of magnitude higher than on the sediments (■ Fig. 9.9). Other species that were found on the turbine structures in the *alpha ventus* offshore windfarm, such as the longspined bullhead, *Taurulus bubalis*, are obligate hard-bottom dwellers and, thus, entirely absent from the sediments.

The seasonal dynamics of the assemblages of large and mobile species are largely unknown. The study showed clear seasonal variations in the abundances of common megafauna species, indicating intensive exchange of individuals between the localised artificial hard substratum and other habitats (■ Fig. 9.9). For example, edible crab is known to migrate tens of kilometres between suitable habitats, e.g. in search for mates or other, still vacant hard-bottom habitats (Bennet & Brown 1983). The observed densities of the mobile demersal megafauna allow for an extrapolation of the population sizes to the total amount of hard substrata that will become available in line with current planning for expansion of the offshore wind sector. These extrapolations predict an increase in population size for some species by several hundred percent (Krone et al. 2013a). The calculation assumes that the introduction of artificial habitats allows for production of additional individuals. As yet, however, it remains to be investigated to what extent the new structures allow for the production of new individuals. Individuals might also be attracted from other, perhaps less favourable habitats (i. e. production versus attraction dilemma; Pickering & Whitmarsh 1997).

9.4 Perspectives

The results show that offshore windfarms will have a substantial effect on the marine benthos. Independent structural variations of the benthic in- and epifauna of the sedimentary seafloor were observed both in the *alpha ventus* area and in a reference area, indicating an influence of the wind turbines and/ or the associated activities (e.g. fishery cessation) on population dynamics of benthic species. Further continuous monitoring will reveal whether the communities of the two areas will further differ or if the observed difference was merely transient.

For decades, both the later windfarm area and the reference area were subject to intense bottom trawling. Since 2008, the windfarm area has been closed to any kind of ship traffic, including fishery vessels, while bottom trawling was still allowed in the reference area. There were no clear signs of recovery of the benthic communities from long-term bottom trawling in the windfarm area. This recovery process appears to take much longer and possibly requires larger untrawled areas to overcome possible edge effects. Depending on sedimentary conditions and the fishing gear used, recovery of benthic habitats after fishery cessation could take up to eight years (e. g. Kaiser et al. 2000) or even longer (Duineveld et al. 2007).

The other – and possibly the major – effect of the introduction of numerous wind turbine constructions into the marine environment is the aggregation and production of marine biota on the submersed structures, resulting in a substantial increase of the structural and functional biodiversity of the benthic

9

system. Tens of tons of fouling organisms and thousands of large mobile demersal animals per windfarm will lead to a drastic shift in the distribution of biomass and thus, in the flow of energy and matter, at least on a local scale. The aggregated organisms interact intensively with each other and with the established fauna of the surrounding sandy seafloor. Accordingly, artificial hard substrata promote the development of complex trophic and competitive interactions within a modified benthic community consisting of both hard- and soft-bottom species. Currently, it is unknown whether these effects will remain locally restricted to single turbines or if they will expand to the entire windfarm area or even beyond resulting in large-scale ecosystem effects.

9.5   Acknowledgements

The sampling campaign in spring 2008 was conducted by BioConsult Schuchardt & Scholle GbR. Thanks go to B. Ebbe for valuable comments on the manuscript.

Literature

Bennet DB & Brown CG (1983). Crab (*Cancer pagurus*) migrations in the English Channel. Journal of the Marine Biological Association of the United Kingdom 63:371–398.

BSH (2007). Standard Investigation of the Impacts of Offshore Wind Turbines on the Marine Environment (StUK3). Bundesamt für Seeschifffahrt und Hydrographie, Hamburg and Rostock, 58 p.

Buchanan JB (1966). The biology of *Echinocardium cordatum* (Echinodermata: Spatangoidea) from different habitats. Journal of the Marine Biological Association of the United Kingdom 46:97–114.

Coates D, Vanaverbeke J, Vincx M (2012). Enrichment of the soft sediment macrobenthos around a gravity based foundation on the Thorntonbank. In: Degraer S, rabant R, Rumes B (eds.) Offshore windfarms in the Belgian part of the North Sea: heading for an understanding of environmental impacts. Royal Belgian Institute of Natural Sciences, Management Unit of the North Sea Mathematical Models, Marine Ecosystem Management Unit, Brussels, pp. 41–54.

Duineveld GCA, Bergman MJN, Lavaleye MSS (2007). Effects of an area closed to fisheries on the composition of the benthic fauna in the southern North Sea. ICES Journal of Marine Science 64: 899–908.

Gili JM & Coma R (1998). Benthic suspension feeders: their paramount role in littoral marine food webs. Trends in Ecology and Evolution 13:316–321.

Green RH (1979). Sampling design and statistical models for environmental biologists. John Wiley & Sons, New York.

Hill T & Lewicki P (2007). STATISTICA: Methods and Applications. StatSoft, Tulsa, OK.

Joschko T, Buck B, Gutow L, Schröder A (2008). Colonization of an artificial hard substrate by *Mytilus edulis* in the German Bight. Marine Biology Research 4:350–360

Kaiser MJ, Ramsay K, Richardson CA, Spencer FE, Brand AR (2000). Chronic fishing disturbance has changed shelf sea benthic community structure. Journal of Animal Ecology 69:494–503.

Krone R, Gutow L, Brey T, Dannheim J, Schröder A (2013a). Mobile demersal megafauna at artificial structures in the German Bight – likely effects of offshore windfarm development. Estuarine, Coastal and Shelf Science 125:1–9.

Krone R, Gutow L, Joschko TJ, Schröder A (2013b). Epifauna dynamics at an offshore foundation – implications of future wind power farming in the North Sea. Marine Environmental Research 85:1–12.

Legendre L & Legendre P (1998). Numerical ecology. Elsevier, Amsterdam.

Lindeboom HJ, Kouwenhoven HJ, Bergman MJN, Bouma S, Brasseur S, Daan R, Fijn RC, de Haan D, Dirksen S, van Hal R, Lambers RHR, Ter Hofstede R, Krijgsveld KL, Leopold M, Scheidat M (2011). Short-term ecological effects of an offshore windfarm in the Dutch coastal zone; a compilation. Environmental Research Letters 6 (035101):13.

Merck T & von Nordheim H (2000). Mögliche Probleme von Offshore-Windenergieanlagen aus Naturschutzsicht. BfN-Skripten 29, 88–98 (2000).

Page HM, Dugan JE, Dugan DS, Richards JB, Hubbard DM (1999). Effects of an offshore oil platform on the distribution and abundances of commercially important crab species. Marine Ecology Progress Series 185:47–57.

Pickering H & Whitmarsh D (1997). Artificial reefs and fisheries exploitation: a review of the 'attraction versus production' debate, the influence of design and its significance for policy. Fisheries Research 31:39–59.

Sadovy Y & Domeier M (2005). Are aggregation-fisheries sustainable? Reef fish fisheries as a case study. Coral Reefs 24:254–262.

Salzwedel H, Rachor E, Gerdes D (1985). Benthic macrofauna communities in the German Bight. Veröffentlichungen Institut für Meeresforschung Bremerhaven 20:199–267.

Underwood AJ (1991). Beyond BACI: experimental designs for detecting human environmental impacts on temporal variations in natural populations. Australian Journal of Marine and Freshwater Research 42:569–587.

Underwood AJ (1992). Beyond BACI: the detection of environmental impact on populations in the real, but variable, world. Journal of Experimental Marine Biology and Ecology 161:145–178.

Underwood AJ (1993). The mechanics of spatially replicated sampling programmes to detect environmental impacts in a variable world. Australian Journal of Ecology 18:99–116.

Wieking G & Kröncke I (2003.) Abundance and growth of the sea urchin *Echinocardium cordatum* in the central North Sea in the late 80 s and 90 s. Senckenbergiana Maritima 32:113–124.

Zintzen V, Massin C, Norro A, Mallefet J (2006). Epifaunal inventory of two shipwrecks from the Belgian continental shelf. Hydrobiologia 555:207–219.

Effects of the *alpha ventus* offshore test site on pelagic fish

Sören Krägefsky

Federal Maritime and Hydrographic Agency,
Federal Ministry for the Environment, Nature Conservation and Nuclear Safety (Eds.)
Ecological Research at the Offshore Windfarm alpha ventus,
DOI 10.1007/978-3-658-02462-8_10, © Springer Fachmedien Wiesbaden 2014

10.1 Introduction

Pelagic fish species like **mackerel** (*Scomber scomber*), **horse mackerel** (*Trachurus trachurus*), **herring** (*Clupea harengus*) and **sprat** (*Sprattus sprattus*) (◻ Fig. 10.1), which dominate the pelagic fish community in wide parts of the German Bight during different seasons of the year, show pronounced seasonal migration behaviour. Throughout the year, they migrate several hundreds of kilometres between their spawning grounds, feeding grounds and overwintering areas. During their migration they react to local changes in their biological and physical marine environment, such as food supply or water temperature. Besides active migration, their distribution is also determined by drift of eggs and larvae in ocean currents. Their stocks consist of different components with different spawning, summering and wintering grounds, showing spatial overlap and mixing during their seasonal migration.

Like most pelagic fish species, all dominant species in the North and Baltic seas show pronounced schooling behaviour. Schools usually consist of individuals of a specific age and thus size class, with differences in spatial distribution between schools of adults and juveniles.

Mackerel, horse mackerel and herring observed at *alpha ventus* each reach sexual maturity at an age of about three years, but differ in maximum age and body length. The potential maximum age for herring, mackerel and horse mackerel is about 25, 30 and 40 years respectively, with maximum lengths of up to 45, 60 and 70 cm. They usually occur at ages below 20 years, showing maximum length of about 30 cm (herring) and 40 cm (mackerel and horse mackerel), respectively. Sprat, in contrast, is a short-living species, reaching sexual maturity after two years and a maximum age of about six years.

All of these species feed on zooplankton, mainly crustaceans, as a main or major food component. Mackerel and horse mackerel eat small fish and other pelagic organisms like cephalopods as a major additional food source.

Fish species like herring and mackerel differ widely in their swimming and hearing capabilities. The mackerel is a fast and powerful swimmer. It does not possess a swim-bladder, and can thus perform fast, vertical movements. However, to avoid sinking and ensure they can breathe, mackerel must swim constantly. Given the lack of a swim-bladder which supports hearing in fish, hearing capability in mackerel is poor. In contrast, specialised structures that link the swim-bladder to the inner-ear of herring make the Atlantic herring a hearing specialist. Thus, effects due to the introduction of underwater structures, such as noise emission during construction and operation of offshore windfarms, may differ widely between pelagic fish species.

Mackerel, horse mackerel, herring and sprat are exposed to high fishing pressure. The North Sea component of the North Atlantic mackerel stock decreased strongly in the 1960 s due to over-fishing and still shows no sign of recovery. The stock of herring in the North Sea collapsed in the 1970 s due to over-exploitation, but recovered in the 1980 s following closure of fisheries activity in the late 1970 s. Again, fishing pressure on juveniles led to a strong decline in spawning stock in the 1990 s. Since the early 2000 s, the stock shows low recruitment from young cohorts, possibly due to climate change-induced impacts on their biological and physical marine environment.

These man-made changes include changes in water temperature and salinity, up to large-scale alterations in the world's ocean current system, as well as changing food supply. Abundance and distribution of zooplankton, which pelagic fish feed upon, is strongly influenced by altered hydrographic conditions as a result of climate change. Thus, effects caused by habitat changes due to construction of offshore windfarms cannot be evaluated separately from effects resulting from these other impacts. Strong fishing pressure on stocks, climate change-induced changes in the marine environment and habitat changes due to windfarm construction may lead to interacting and possible mutually reinforcing effects.

Due to extensive migration behaviour, there is a sequence of impacts along drift and migration paths of juveniles and adults, and within the separate spawning, feeding and overwintering grounds, causing short-term effects (e.g. on changing migration behaviour) and long-term effects (e.g. on reproduction). Moreover, local occurrence of pelagic fish is very patchy due to their schooling behaviour. A survey of the impacts of

Fig. 10.1 Typical pelagic fish species: (a) mackerel (*scomber scomber*), (b) horse mackerel (*Trachurus trachurus*), (c) herring (*Clupea harengus*), (d) sprat (*Sprattus sprattus*) (photo: (a-d) Sven Gust).

offshore windfarms on pelagic fish in the existing context of fishery and climate change-induced impacts must thus address processes on very different temporal and spatial scales. This demands appropriate survey methods.

At present, all studies on the impact of offshore windfarms on pelagic fish are restricted to a narrow survey area of a localised windfarm and surrounding reference areas. The same applies for investigations at the *alpha ventus* offshore windfarm. These include surveys on the distribution of pelagic fish in the *alpha ventus* windfarm area and the surrounding area based on hydroacoustic measurements, addressing immediate attraction and repulsion effects of offshore windfarms. Attraction or repulsion may result from a set of effects, which include direct changes in the habitat (electromagnetic fields (EMF), noise, structure) and indirect changes (food

and predators). The effect of changes in the food supply was addressed as a single effect by analysis of the gut content of pelagic fish in and outside the *alpha ventus* area.

The study was performed over the period from August 2008 to April 2012.The survey covers a period before construction, during construction and during operation of the windfarm, in different seasons (spring, summer and autumn), and covering the succession in occurrence of the pelagic fish species in the study area due to their seasonal migration behaviour. The study further includes an evaluation of the survey methods, including development of a stationary measuring system for hydroacoustic long-term monitoring of fish. Such long-term monitoring should be part of an overarching (cross-windfarm) study on the effects of offshore windfarms, as discussed below.

10.1.1 **Possible impacts**

Construction and operation of all windfarms so far approved and those applied for would cause direct and indirect changes to the pelagic fish habitat in the North and Baltic seas. Direct changes in habitat are due, for example, to the construction of underwater structures, EMF emissions in immediate proximity to underwater power cables, and noise emitted into the water. Noise intensities are high during wind-farm construction and are caused by pile driving, while noise emission is much lower during wind-farm operation. Indirect changes in the pelagic fish habitat stem, for example, from changes in local food supply caused by the settlement of hard-bottom species that take benefit from the introduction of underwater structures. There could be a shelter effect due to the exclusion of fishery activities inside the windfarm.

For most hard-substrate-associated animals, such as some crustaceans or mussels, possible positive effects of offshore windfarms can be deduced directly from their settlement success. However, it is much more difficult to deduce positive or negative effects on most demersal and all pelagic fish species. This is due to their extensive migration behaviour. Assessment of the impact of offshore windfarms so far is based on laboratory studies, on the few available field studies and on observations on separate effects of EMF, such as noise or introduction of structures in the ocean.

Electromagnetic fields

Reception of magnetic fields has been examined in studies with several diadromous (migrating between fresh and salt water) and a few oceanodromous fish (migrating within salt water only) (Gill 2010). For a number of fish species, sensitivity and reaction to magnetic fields has been proven. These included quite diverse species such as plaice (Metcalfe et al. 1993) and yellowfin tuna (Walker et al. 1984). However, only in a few field studies were the effects of EMF caused by underwater cables directly addressed, showing little or no significant instantaneous change in orientation and swimming behaviour.

Magnetic fields of a certain strength may cause physiological reactions in fish and are shown, for example, to have an effect on embryonic development

in trout (Formicki et al. 1998). Exposure of early stages of the Japanese eel to strong magnetic fields can cause change in orientation behaviour at adult stage (Nishi & Kawamura 2005). Species showing such an effect would show a bias in migration behaviour without being exposed to EMF at the time.

The magnitude of EMF a fish is exposed to in the vicinity of a power cable decreases rapidly as the inverse square of the distance from the cable. Most likely, only in the direct vicinity of a cable are fish exposed to EMF to such an extent that it has significant effects.

Underwater noise

Fish perceive and produce sound. Environmental sound can deliver information on abiotic and biotic surroundings, signalling, for example, the presence of predators and prey. Fish themselves produce sound for the purpose of predator deterrent and communication, serving, for instance, as a signal in the mating behaviour of some species (Kasumyan 2008). Fish can perceive sound through perception of particle motion and pressure variations. The particle motion is detected by the inner ear or the lateral line organ of fish. However, perception of the pressure component of sound requires the presence of an air-filled swim-bladder. The swim-bladder serves as an aiding structure by which sound pressure is translated to mechanical stimulus perceived with otoliths, a kind of biological accelerometer located in the inner ears of fish. Hearing ability of species like mackerel, which do not feature a swim-bladder, is low compared to other species. Special morphological connections between the swim-bladder and the inner ear allow specialised fish to perceive sound in a relatively broad frequency range and at low intensity. Among such specialists are the Clupeoids (e.g. herrings and sardines).

Noise influence can be divided into different grades of impact reached in certain zones around the noise source (Richardson et al. 1998). The threshold of perception is reached at a certain distance from the source and depends on sound characteristics and species-specific hearing abilities. It therefore varies significantly between mackerel and herring. Increasing sound intensity towards the noise source leads to masking of signals whose perception can serve orientation or communication.

Near the source it may cause temporary or permanent hearing loss, or, at very high intensities in close proximity, even more severe injury and death.

In the construction of a windfarm, noise is emitted by a number of sources. The most important is pile driving when building the foundations for wind turbines and the transformer station. Pile driving generates a sequence of short acoustic pulses with very high sound intensity, and most energy occurs in a frequency band within the main perception range of fish (Itap 2011). Based on hearing sensitivities determined in laboratory studies, herring and cod are likely to perceive noise generated by pile driving from a distance of more than 80 km. However, there are only a few field studies on the reaction and impact of pile-driving generated noise on single fish species. A current mesocosm study shows significant changes in behavioural reactions of sole (*Solea solea*) and cod (*Gadus morhua*) occurring at relatively low sound intensities (Mueller-Blenkle et al. 2010). Most field studies so far are studies on injuring effects on encaged fish in close proximity to the noise source. These studies show contradicting results, including inner and outer damage to body tissue depending on the distance from the source (Caltrans 2001), while some studies show no effect (Oestman & Earle 2012).

During operation of a windfarm, noise is mainly emitted by generators and gear. The sound frequency range of this noise is within the general perception range of fish (Itap 2011). The noise intensity is fairly low compared to the noise intensity from pile-driving, and cannot cause immediate physical harm to fish. The estimated perception distance of windfarms during operation is below 10 m for fish without a swim-bladder and below 1 km for fish with normal hearing capabilities. Hearing specialists may sense operating windfarms at a distance of up to 10 km (Andersson 2011). Only in the direct vicinity of a wind turbine is the noise level likely to be sufficient to induce a behavioural reaction in some fish species. The effects of low level but long-term noise exposure in the vicinity of a wind turbine, e.g. whether it is a stressor for some fish, are unknown.

Structure

There are numerous observations of fish aggregation around floating or stationary structures in the sea (e.g. Kingsford 1993, Dempster & Taquet 2004). Floating structures include floating organisms like jellyfish or algae as well as natural and artificial flotsam such as garbage, and also objects deployed for fisheries purpose to attract fish. Underwater structures like shipwrecks or sites such as oil platforms are colonised by hard-substrate-associated organisms and fish (Zintzen et al. 2006, Helvey 2002). In this function they are referred to as secondary artificial reefs (e.g. Love et al. 1999).

The ability of the local fish population to increase, however, relies only to a minor extent on local production (Bohnsack 1989). Any increase in the abundance of migratory pelagic fish would be due to a re-distribution of fish due to attraction. The reasons for such attraction are contentious, and based largely on assumptions. The reasons most frequently assumed are usage of underwater structures a) as a protection and survival area, b) as a feeding habitat for fish feeding on hard-substrate-associated species, and c) as a hunting area for predatory fish feeding on smaller fish (Bohnsack 1989). However, the fundamental assumption that observed aggregation behaviour is of actual benefit does not necessarily hold. Aggregation can take place because of visual (e.g. light and shade) and non-visual signals that might otherwise be specific signals serving as a cue for behavioural response gaining advantage in another situation. Moreover, in some cases, aggregation behaviour might be even a disadvantage for some fish species – for example, if the energetic effort of the behaviour exceeds the gain. There are even more indications that aggregation behaviour around artificial underwater structures can adversely affect the temporal and spatial coherence of migratory fish populations (Wang et al. 2012).

Food supply

The construction of offshore windfarms leads to colonisation of underwater structures by hard-substrate-associated species. These are potential food for demersal and pelagic fish. So far only a few studies are available on the feeding behaviour of fish in offshore windfarms. And only one published study looks at a pelagic species, the horse mackerel (Derweduwen et al. 2012). The studies include visual observations (by scuba divers) and stomach content analysis. May (2005) observed schools of

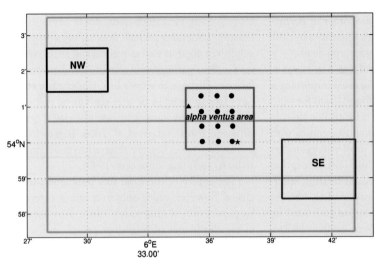

□ **Fig. 10.2** Mesoscale survey area (green transect lines) around the *alpha ventus* test site (red frame). Black dots mark the location of single wind turbines, the black triangle marks the location of FINO1 and the black star the location of the transformer station. The reference areas are framed in black (NW) and blue (SE).

whiting (*Merlangius merlangus*) feeding on the amphipoda *Jassa falcata,* which was densely colonising the underwater structures of wind turbines. Stomach content analyses revealed differences in the diet composition of individuals from some fish species caught at different sampling sites. Likewise, some species showed a higher grade of gastric filling in the vicinity of the windfarm compared to the reference areas (Derweduwen et al. 2012). Whether these differences are based on a difference in local food supply is vague. Only for a few benthic (dab, *Limanda limanda*) and bentho-pelagic fish species (pouting, *Trisopterus luscus*) has preferential feeding on hard-substrate-associated prey items been ascertained (Derweduwen et al. 2012, Reubens et al. 2011). The diet of the examined horse mackerel consisted of pelagic crustaceans. Hard-substrate-associated prey items constituted an insignificant source of food (Derweduwen et al. 2012).

10.2   Methods

Pelagic fish are patchily distributed. Their stocks cannot be assessed with net catches. Because ship-based hydroacoustic surveys allow for wide but nevertheless temporally and spatially high-resolution measurement of fish distribution, they are an international standard for assessment of local fish populations as well as estimation of fish stocks in large scale sea areas (▶ Information box: *Hydroacoustic survey methods*).

The study described in this section covered ship-based surveys in autumn 2008 (pre-study) and each spring, summer and autumn in 2009, 2010 and 2011. Multi-day ship-based hydroacoustic surveys were carried out inside and outside the windfarm before and during construction and operation of *alpha ventus.* The measurements were made with a hull-mounted multi-frequency echosounder on the HEINCKE research vessel along transects within *alpha ventus* and two reference areas in the surrounding area with similar seabed characteristics and depth, and within an area of 200 km² with *alpha ventus* as its centre (□ Fig. 10.2). Additional net catches with a pelagic trawl (PTN) allowed identification of fish species composition and size distribution (□ Fig. 10.3).

Given an identical species and size composition of fish in the whole study area, hydroacoustic measurements, which can be expressed as an areal backscattering coefficient – nautical area scattering coefficient (NASC) – are directly proportional to the abundance of pelagic fish. Due to the very patchy distribution of such fish, however, the mean or median NASC may over or underestimate the real abundance of pelagic fish in a given local area. For this reason, comparison of the fish abundance inside and outside *alpha ventus* was performed with the help of a statistical model (GLM) that additionally takes into account the possibility of a preferred site of residence for fish in areas of certain depths.

To examine the influence of windfarms on the feeding behaviour of pelagic fish, the stomach content of horse mackerel and mackerel was analysed. These two species have a similar diet but show species-specific differences – for instance in swimming behaviour – that can result in different possibilities for the use of underwater structures be-

longing to wind turbines as a feeding habitat. The stomach content of about 400 fish was examined between 2008 and 2011. This included the stomachs of 280 fish caught during operation of *alpha ventus*. Fish were caught inside and outside the windfarm

Hydroacoustic survey methods

Active hydroacoustic measurements allow a survey of marine organism distribution in the size range of small macrozooplankton to large fish, and with very high temporal and spatial resolution. This cannot be achieved with any other survey method. Echosounder measurements are routinely used for biomass stock estimates. In fisheries science, hydroacoustic surveys are defined as the standard stock assessment tool for purpose of fisheries management.

In hydroacoustic surveys, sound at high sound frequencies is transmitted into the water at a known (low) intensity. This sound is backscattered by pelagic organisms like zooplankton and fish, and is received back by the echosounder transducer. The strength of backscattering from single organisms is dependent on their shape, size and material properties, and the sound frequency used. Zooplankton, for example, are weak backscattering organisms as they have material properties similar to the surrounding water, while fish with a swimbladder produce relatively strong backscattering signals.

The specific backscattering characteristics of marine organisms, including the characteristic shape of schools of different fish species in an echogram (a visualisation of measured sound backscattering within the water column), are used for identification of fish and fish species. After excluding echoes, such as from zooplankton, by applying a threshold or by exclusion of regions, the measured backscattering strength in a given area can be used as a measure of fish abundance.

Hydroacoustic measurements can be performed as a ship-based survey along transects or with a stationary echosounder, both in a vertically or horizontally aligned measuring mode. Horizontal measurements were used to survey fish abundance around single wind turbines. These measurements span a horizontal distance from the research vessel to the wind turbine while the ship

keeps at its safety distance from the turbine. However, the sound beam spreads along its path and thus samples different depth intervals with distance from the echosounder. Where the depth distribution of fish is non-homogeneous, this causes a biased abundance estimate of the distribution of fish around a wind turbine. Most pelagic fish, including species like herring, mackerel and horse mackerel, show a preferential residence depth in the water column, usually changing from day to night. Thus, in most cases, horizontal measurements are an inappropriate method to survey fish abundance around wind turbines.

Ship-based hydroacoustic surveys provide large spatial coverage of fish stock assessments and thus are indispensable when surveying the impact of offshore windfarms. However, the resulting measurement is a snapshot. The temporal coverage that can be achieved by repeat surveys is restricted by survey effort, high costs and dependency on appropriate weather and sea conditions. Stationary measuring systems allow for reliable survey of fish abundance in direct vicinity of a wind turbine and, furthermore, for long-term measurements, thus potentially providing seasonal and multi-year coverage during harsh weather periods.

Power consumption of a modern multi-frequency echosounder during operation is in the order of a light bulb, if electronic devices (e.g. computers) with low power consumption are used. However, this energy demand is still a major restriction to the achievable temporal coverage for measurements with stationary systems without a cable link for power. As part of the research work and the evaluation of survey methods in *alpha ventus*, an innovative stationary hydroacoustic measuring system was developed with the objective of significantly reducing energy consumption during deployment, to allow for easy deployment and recovery, and to provide for cable-linked and battery-driven operation.

■ **Fig. 10.3** Illustration of applied monitoring methods. Hydroacoustic survey: Sound is emitted within a narrow sound beam by a ship-mounted (1) or stationary echosounder (3). Echoes caused by backscattering from fish are received, allowing for abundance estimates. (3) shows the newly developed stationary hydroacoustic measuring system. (2) shows the net catches with the pelagic trawl and (4) the rod catches for the analysis of the stomach content.

area and in the vicinity of the FINO1 research platform.

10.3 Results and discussion

Net catches in spring and autumn were dominated by herring and sprat (■ Fig. 10.4). European anchovy (*Engraulis encrasicolus*) and sardine (*Sardina pilchardus*) occurred in significant numbers. Like herring and sprat, these belong to the Clupeoids but are warm-water species whose abundance in the North Sea has increased with rising mean water temperatures (Alheit et al. 2012). During summer, mackerel was found to be an important or dominant pelagic fish species in *alpha ventus* (■ Fig. 10.4). These seasonal changes in composition of the net catches reflect the seasonal migration behaviour of pelagic fish. As in other areas of the North Sea, herring, sprat, mackerel and horse mackerel dominated the catches during different times of the year. They are the most relevant pelagic fish species for commercial fishery in the German Bight. Their relative shares in species

composition and their ages, should, however, differ between windfarms with increasing distance from the shore.

The composition of pelagic fish species inside and outside *alpha ventus* is strongly congruent (■ Fig. 10.4). The similarity between net catches (species composition by weight) taken inside *alpha ventus* and the reference areas can be expressed by an index, where the value 0 signifies total similarity and 1 dissimilarity. During the entire study, the index was mostly below 0.05 and showed a maximum of 0.17, suggesting identical composition inside and outside the windfarm.

For the pre-construction phase, the hydroacoustic survey (spring 2009) shows no unequal distribution of pelagic fish in relation to the *alpha ventus* area. Measurements carried out during construction of the windfarm show low abundance of pelagic fish in the *alpha ventus* area compared to the surrounding area. Determining NASC in relationship to the area concerned shows a 50 % (summer 2009) and 40 % (autumn 2009) decrease in fish abundance, respectively, inside *alpha ventus* relative to its surroundings (■ Fig. 10.5).

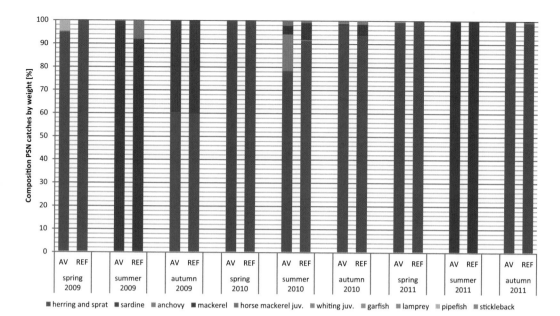

Fig. 10.4 Relative composition of pelagic fish species in net catches (pelagic trawl) by weight during the surveys in spring, summer and autumn in 2009, 2010 and 2011 inside the *alpha ventus* area (AV) and the reference areas (REF).

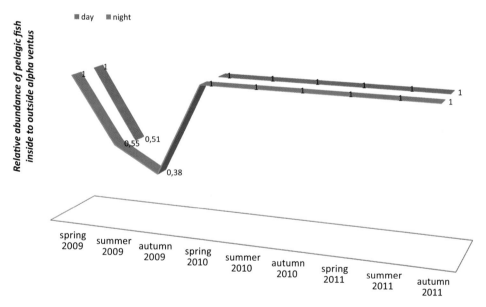

Fig. 10.5 Relative abundance of pelagic fish inside *alpha ventus* compared to the surrounding area before construction (spring 2009), during construction (summer and autumn 2009) and during operation (2010, 2011). There is insufficient survey coverage for a proper night-time assessment inside the construction area in autumn 2009.

Known sensitivities and hearing capabilities of fish and high noise intensities occurring during pile driving suggest that fish will be scared away. Fish are able to perceive the sound of ships and can demonstrate flight reaction to approaching vessels (Mitson & Knudsen 2003). Additional but less significant scaring due to other construction activities can thus be assumed.

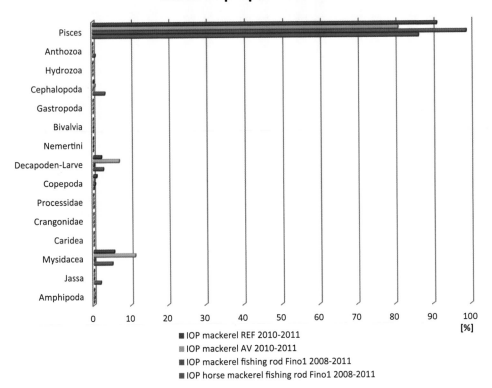

index of preponderance

Y-axis categories (top to bottom): Pisces, Anthozoa, Hydrozoa, Cephalopoda, Gastropoda, Bivalvia, Nemertini, Decapoden-Larve, Copepoda, Processidae, Crangonidae, Caridea, Mysidacea, Jassa, Amphipoda

X-axis: 0, 10, 20, 30, 40, 50, 60, 70, 80, 90, 100 [%]

- ■ IOP mackerel REF 2010-2011
- ■ IOP mackerel AV 2010-2011
- ■ IOP mackerel fishing rod Fino1 2008-2011
- ■ IOP horse mackerel fishing rod Fino1 2008-2011

▣ **Fig. 10.6** Index of preponderance (food index weighted by the weight and occurrence of food items in fish stomachs), that is, the relative importance of different food for mackerel and horse mackerel caught in the *alpha ventus* area and reference areas.

During operation of the windfarm, the hydroacoustic surveys conducted in spring, summer and autumn revealed no significant difference in abundance of pelagic fish inside or outside *alpha ventus* that could be attributed to an impact from the windfarm. Presently, however, the close vicinity of a wind turbine is a blind zone for the assessment of the fish stock inside the windfarm.

Dependent on their swimming behaviour, feeding behaviour or hearing capabilities, the different fish species may show preferential residence in close proximity to wind turbines due to attraction by underwater structures or food, or they may avoid the zone due to noise emission during operation. Thus, because of its unavoidable exclusion of the zone closest to the wind turbines, the ship-based assessment might under or overestimate the total fish stock in the windfarm in the case of species that show attraction or repulsion to the wind

turbines. This blind zone problem can be solved with the help of a stationary long-term measurement system.

Analysis revealed that hard-substrate-associated organisms are of minor relevance in the diet of horse mackerel and mackerel, or for those fish caught in the direct vicinity of underwater structures (▣ Fig. 10.6). Their food consisted nearly exclusively of fish, as the major food source, and zooplankton. As assumed in terms of the natural food spectrum, horse mackerel and mackerel do not derive any direct advantage from windfarm-related introduction of hard-substrate-associated organisms. The stomachs of mackerel caught inside the windfarm area ($n = 102$) were significantly less full than individuals caught in the reference areas ($n = 97$). Including mackerel caught with the help of fishing rods in the close vicinity of FINO1 ($n = 50$), the average fullness was 57 % (54 % PTN catches) of the average stomach fullness determined for reference areas.

Stomach fullness alone does not allow conclusions on the total food gathered in a certain period of time (e.g. per day). Feeding on zooplankton (low weight) occurs continuously, while in contrast other fish (high weight) are preyed sporadically. In view of strong similarity in diet composition of mackerel caught inside and outside *alpha ventus*, however, the lesser stomach fullness invites the assumption that mackerel inside the windfarm area showed less food consumption.

There are several possible reasons for decreased food gathering. Feeding behaviour may be disturbed or there may be differences in abundance and composition of prey. Noise has a potential impact on feeding behaviour, but this is rarely studied in pelagic fish. A study on the behaviour of the three-spined stickleback (*Gasterosteus aculeatus*) revealed reduced ability to recognise food, errors in food handling and less efficiency (Purser & Radford 2011). Underwater structures can also influence the feeding patterns of constantly swimming mackerel. Disturbance of swimming behaviour and thus of food gathering is nonetheless only likely in the close vicinity of the underwater structures. To further address causes of an apparent decrease in food consumption, however, it is necessary to synoptically survey the distribution and species composition of the food organisms of pelagic fish. These are mainly zooplankton, small fish and other pelagic organisms. Their distribution correlates with hydrographic parameters, is subject to coincidental changes and is possibly influenced by the presence of a windfarm (e.g. occurrence of larvae of hard-substrate-associated species).

10.4 Perspectives

The migration behaviour of pelagic fish makes it impossible to survey long-term effects of offshore windfarms at the level of a single windfarm. With further construction of offshore windfarms in the German Bight, any survey conducted must take a cross-windfarm approach, and must include large-scale surveys on distribution and stock size of fish

species as well as long-term measurements using stationary measuring systems in selected windfarm and reference areas. These long-term measurements allow for reliable abundance estimates, including for fish in close proximity to wind turbines, and make it possible to discriminate between long-term trends and short-term fluctuations.

Large-scale surveys and stationary measurements must take the form of synoptic surveys covering local abundance and composition of food organisms and hydrographic parameters to be able, for instance, to address the impact on feeding behaviour of fish species. Such synoptic surveys can be performed using a combination, among other measuring techniques, of optical and hydroacoustic measurements. In the case of ship-based surveys, this can be done with the aid of a towed vehicle equipped with a set of sensors. Synoptic measuring techniques that allow synoptic surveying are increasingly used during standard stock assessment surveys. Long-term measurements can be performed with an enhanced version of the above-mentioned stationary hydroacoustic measuring system, including, among other sensors, an optical (image-based) measuring system for survey of pelagic food organisms.

Literature

Alheit J, Pohlmann T, Casini M, Greve W, Hinrichs R, Mathis M, Vorberg R, Wagner, C (2012). Climate variability drives anchovies and sardines into the North and Baltic Seas. Progress in Oceanography 96:128–139.

Andersson MH (2011). Offshore windfarms – ecological effects of noise and habitat alteration on fish. PhD thesis, Department of Zoology, Stockholm University.

Bohnsack JA (1989). Are High Densities of Fishes at Artificial Reefs the Result of Habitat Limitation or Behavioral Preference? Bulletin of Marine Science 44:631–645.

Caltrans (2001). Pile Installation Demonstration Project. Fisheries Impact Assessment, Caltrans Contract 04A0148. San Francisco – Oakland Bay Bridge East Span Seismic Safety Project.

Dempster T & Taquet M (2004). Fish aggregation device (FAD) research: gaps in current knowledge and future directions for ecological studies. Reviews in Fish Biology and Fisheries 14:21–42.

Derweduwen J, Vandendriessche S, Willems T, Hostens K (2012). The diet of demersal and semi-pelagic fish in the Thorntonbank windfarm: tracing changes using stomach analyses data. In: Degraer S, Brabant R, Rumes B (Eds.) Offshore windfarms in the Belgian part of the North Sea., pp. 73–84.

Formicki K & Winnicki A (1998). Reactions of fish embryos and larvae to constant magnetic fields. Italian Journal of Zoology 65:479–482.

Gill A & Bartlett M (2010). Literature review on the potential effects of electromagnetic fields and subsea noise from marine renewable energy developments on Atlantic salmon, sea trout and European eel. Scottish Natural Heritage Commissioned Report 401.

Helvey M (2002). Are southern California oil and gas platforms essential fish habitat? ICES Journal of Marine Science 59:266–271.

ITAP (2011). Messungen von Unterwasserschall beim Bau der Windenergieanlagen im Offshore-Testfeld alpha ventus. Abschlussbericht zum Monitoring nach StUK 3 in der Bauphase, pp. 1–48.

Kasumyan AO (2008). Sounds and sound production in fishes. Journal of Ichthyology 48:981–1030.

Kingsford MJ (1993). Biotic and abiotic structure in the pelagic environment: importance to small fishes. Bulletin of Marine Science 53:393–415.

Love M, Caselle J, Snookm L (1999). Fish assemblages on mussel mounds surrounding seven oil platforms in the Santa Barbara Channel and Santa Maria Basin. Bulletin of Marine Science 65:497–513.

May J (2005). Post-construction results from the North Hoyle offshore windfarm. Paper for the Copenhagen offshore wind international conference, Project Management Support Services Ltd., pp. 1–10.

Metcalfe J, Holford B, Arnold G (1993). Orientation of plaice (*Pleuronectes platessa*) in the open sea: evidence for the use of external directional clues. Marine Biology 117:559–566.

Mitson RB & Knudsen HP (2003). Causes and effects of underwater noise on fish abundance estimation. Aquatic Living Resources 16:255–263.

Mueller-Blenkle C, McGregor P, Gill A, Andersson M M, Metcalfe J, Bendall V, Sigray P, Wood D, Thomsen F (2010). Effects of Pile-driving Noise on the Behaviour of Marine Fish. COWRIE Ref: Fish 06-08, Technical Report 31st March 2010.

Nishi T & Kawamura G (2005). *Anguilla japonica* is already magnetosensitive at the glass eel phase. Journal of Fish Biology 67:1213–1224.

Oestman R & Earle C (2012.) Effects of Pile-Driving Noise on *Oncorhynchus mykiss* (Steelhead Trout). In Popper, AN, Hawkins A (Eds.) Advances in Experimental Medicine and Biology. Springer, New York, pp. 263–265.

Purser J & Radford A (2011). Acoustic Noise Induces Attention Shifts and Reduces Foraging Performance in Three-Spined Sticklebacks (*Gasterosteus aculeatus*). PLoS ONE 6(2) e17478. doi:10.1371/journal.pone.0017478.

Reubens J, Degraer S, Vincx, M (2011). Aggregation and feeding behaviour of pouting (*Trisopterus luscus*) at wind turbines in the Belgian part of the North Sea. Fisheries Research 108:223–227.

Richardson WJ, Greene CR Jr, Malme CI, Thomson DH (1998). Marine Mammals and Noise. Academic Press.

Walker M, Kirschvink J, Chang S, Dizon A (1984). A candidate magnetic sense organ in the Yellowfin tuna *Thunnus albacares*. Science 224:751–753.

Wang X, Xu L, Chen Y, Zhu G, Tian S, Zhu J (2012). Impacts of fish aggregation devices on size structures of skipack tuna *Katsuwonus pelamis*. Aquatic Ecology 46(3): 342–352.

Zintzen V, Massin C, Norro A, Mallefet J (2006). Epifaunal Inventory of Two Shipwrecks from the Belgian Continental Shelf. Hydrobiologia 555:207–219.

Effects of the *alpha ventus* offshore test site on distribution patterns, behaviour and flight heights of seabirds

*Bettina Mendel, Jana Kotzerka, Julia Sommerfeld, Henriette Schwemmer,
Nicole Sonntag, Stefan Garthe*

Federal Maritime and Hydrographic Agency,
Federal Ministry for the Environment, Nature Conservation and Nuclear Safety (Eds.)
Ecological Research at the Offshore Windfarm alpha ventus,
DOI 10.1007/978-3-658-02462-8_11, © Springer Fachmedien Wiesbaden 2014

11.1   Introduction

The North Sea and its adjacent waters are an area of world-wide importance for seabirds. This holds true both for birds breeding along all coasts and for birds using the area during migration and wintering. Numbers during breeding and in the non-breeding period are of international importance (Skov et al. 1995). Within the German sector of the North Sea, 28 species occur regularly. The most numerous species in the German Exclusive Economic Zone (EEZ) are northern fulmar (*Fulmarus glacialis*), lesser black-backed gull (*Larus fuscus*) and black-legged kittiwake (*Rissa tridactyla*) in summer and common guillemot (*Uria aalge*), herring gull (*Larus argentatus*) and black-legged kittiwake in winter (Garthe et al. 2007).

With regard to the establishment of offshore windfarms, it becomes necessary to investigate the possible effects of these installations on seabirds. Thus, an environmental impact assessment (EIA) must be conducted prior to approval of a windfarm site, during construction and subsequently during operation. The EIA must be conducted according to BSH Standard for Environmental Impact Assessment (StUK). To assess possible effects on seabirds, ship-based surveys for offshore windfarms have been conducted since 2000 and aerial surveys commenced in 2001.

To evaluate and develop the methods applied under the StUK3 standard, and to assess possible effects of offshore windfarms on marine wildlife, two studies on seabirds were set up within the framework of the StUKplus project. The TEST-BIRD project investigated the effects of the *alpha ventus* windfarm on seabirds. In that project, distribution patterns of seabirds were determined and detailed observations made on the behaviour and flight heights of individual birds in and around the windfarm area. A recently developed digital method for aerial surveys was also tested in the course of this project.

In the second project, data on seabird distributions from all EIA studies conducted prior to the construction of offshore windfarms, during the construction period and during the operational period, was combined with the long-time dataset from various research projects conducted by Research and Technology Centre (FTZ) in Büsum. This allowed improvement of the combined informative value of the various data sets collected for different windfarm areas. It also allowed calibration and quality control of the survey methods in order to improve data quality and perform combined data analyses.

Data was used from the combined database on seabird abundance and distribution in order to make predictions and draw conclusions about selected seabird species in relation to *alpha ventus*. It also looks at possible differences in seabird densities prior to the windfarm's construction and during the operational phase. Finally, using the TESTBIRD project as a basis, it outlines the findings regarding the behaviour and flight heights of selected seabird species.

11.1.1   Key species

Analyses for seabird distribution, habitat modelling and observations of behaviour and flight heights were made for various seabird species that are either abundant or of particular interest in the investigated areas. This section provides examples of the maWin seabird species that occur in the *alpha ventus* area (for references, see Mendel et al. 2008; (▣ Fig. 11.1, ▣ Fig. 11.2).

In German waters, the **divers** family (*Gavia sp.*) is represented by the **red-throated diver** (*Gavia stellata*) and the **black-throated diver** (*Gavia arctica*). Both species show similar breeding and wintering distributions. However, the red-throated diver is much more numerous in German North Sea waters, with more than 90 % of all observed and identified divers. Both diver species breed in the Arctic and boreal regions of Eurasia and North America. They nest usually on shallow and swampy lakes but are mostly found in marine waters outside of the breeding season. In Germany, they are found during their migration and wintering periods, from autumn to spring – predominantly in the North and Baltic seas. Depending on the season, divers are common along the coast and also in sea areas up to 100 km from the coast.

Divers are opportunistic feeders, but benthopelagic swarming fish species like herring (*Clupea harengus*) and cod (*Gadus morhua*) form an impor-

◘ **Fig. 11.1** Key species: (a) Black-throated diver (*Gavia artica*), (b) red-throated diver (*Gavia stellata*) (photo: (a) Sven-Erik-Arndt, (b) Mathias Putze).

tant part of their diet. They capture their prey by pursuit diving using their feet for propulsion. During the breeding season they also use amphibians or small invertebrates as food sources.

Northern gannets (*Sula bassana*) are large seabirds with a characteristic sharp pointed head and bill. Adult birds are white with a yellowish head and black wing tips. It is principally a migratory species with lots of individuals from European breeding colonies spending their winter time along the coast of West Africa, while other birds stay near the breeding colonies in the North Sea or move only shorter distances southwards. The northern gannet breeds in Germany solely on the island of Heligoland, where breeding bird numbers increased steadily in recent years.

Their main food sources are schooling fish species like sandeels (*Ammodytes* sp.), Atlantic mackerel (*Scomber scombrus*), herring or sprat (*Sprattus sprattus*) which they usually capture by deep plunge or even pursuit diving using their wings for propulsion. Northern gannets forage mainly on continental shelf areas at distances up to several hundred kilometres from their colonies.

The **lesser black-backed gull** (*Larus fuscus*) is a medium-sized gull with dark grey to black upper wings and black and yellow legs. It is a migratory species that usually occurs in German waters during the breeding season from March to early October. Outside the breeding period it migrates south to western Europe and Northwest Africa. Breeding numbers increased significantly in Germany in re-

cent years. The species now belongs to one of the most numerous coastal breeding birds in the country. By far the largest colonies are found along the German North Sea coast, while numbers of breeding lesser black-backed gulls on the Baltic Sea coast are still very low.

Lesser black-backed gulls are primarily offshore feeders that forage up to 80 km from their colonies on fish and marine invertebrates (e.g. swimming crabs *Portunidae*). Discards from fishing boats are also frequently used as food source. However, during recent years many lesser black-backed gulls have fed abundantly inland, on terrestrial food such as insects and earthworms.

The **common guillemot** (*Uria aalge*) belongs to the family of auks (*Alcidae*) and is characteristically black on its upper parts and white on its belly. It is a pelagic seabird that only goes on land during the breeding season. Common guillemots nest in large colonies on steep cliffs and breed in Germany solely on the island of Heligoland.

They feed predominantly on pelagic schooling fish like sandeels (*Ammodytes* sp.) or clupeids (herring and sprat) by means of pursuit diving and using their wings for propulsion. Guillemots are common in German waters year-round and reach their highest densities during the winter time when parts of other European populations migrate into German North Sea and Baltic Sea waters.

The **razorbill** (*Alca torda*) is very alike the common guillemot in its appearance, and also in its distribution in German waters and feeding behaviour,

🖸 **Fig. 11.2** Key species: (a) northern gannet (*Sula bassana*), (b) lesser black-backed gull (*Larus fuscus*), (c) common guillemot (*Uria aalge*), (d) razorbill (*Alca torda*), (e) black-legged kittiwake (*Rissa tridactyla*), (f) little gull (*Hydrocoelus minutus*) (photo: (a) Stefan Garthe, (b-f) Mathias Putze).

although it is generally found in slightly more coastal areas and in much lower numbers. Only few pairs breed in German waters, on the island of Helgoland.

The **black-legged kittiwake** (*Rissa tridactyla*) is a small gull with a yellow bill, black legs and a grey-white plumage. Kittiwakes are pelagic birds that only come on land during the breeding season. Birds breed in Germany only on the island of Helgoland. Kittiwakes are restricted to prey that either

occurs at the surface or within the first few metres of the water column. During plunge-diving, birds can reach depth of 1–2 metres. The diet of Kittiwakes consists predominantly of small, shoaling pelagic fish or marine invertebrates. Kittiwakes also forage behind fishing vessels.

Kittiwakes are agile flyers and swimmers. Flight height and speed strongly depend on wind velocity. During summer and autumn, birds are abundant in

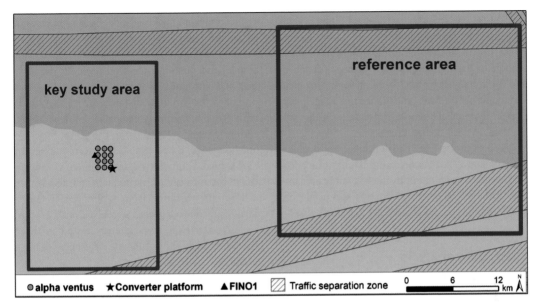

■ **Fig. 11.3** Key study area around *alpha ventus* and the reference area.

the vicinity of Heligoland, and fairly evenly distributed within the EEZ during winter.

The **little gull** (*Hydrocoloeus minutus*) is a small, light bird. Its wings are coloured light grey above and blackish-grey below. In their breeding plumage, little gulls are black hooded. This species only occurs as a migrant in German waters during its non-breeding season. Little gulls are often observed in flocks, showing a characteristic migration and feeding pattern. Birds mainly forage in flight and also from the surface while swimming. Little is known about their diet during the migrating or wintering period. Probable food items include zooplanktonic organisms, insects and little fish.

11.2 Methods

11.2.1 Survey design

In the course of the TESTBIRD project, two different areas were surveyed. One was the area in the vicinity of *alpha ventus* (key study area), and a second, further east, was used as a reference area, where no windfarm related disturbances were expected (■ Fig. 11.3). For the purpose of investigating distribution patterns, only data from the key study area was considered. For analysis of behav-

iour and flight heights, data was taken from both areas.

Data on seabird distributions was based both on the data gathered in the TESTBIRD project and on data from pre- and post-construction EIA studies. These surveys were conducted as ship-based and aerial surveys.

Both survey types were transect counts to determine distributional patterns of seabirds. They were carried out according to international standards. Aerial surveys were conducted according to the methods described in Pihl & Frikke (1992), Noer et al. (2000) and Diederichs et al. (2002). Ship-based surveys additionally allowed observation of seabird behaviour and were conducted in accordance with the standards of the European Seabirds at Sea Specialist Group (Tasker et al. 1984, Garthe et al. 2002). The transect counts recorded all birds swimming or flying inside a pre-defined transect width (ship-based surveys: 300 m, aerial survey: 397 m). If time allowed, birds outside these transect strips were also registered. The area surveyed was calculated from the distance travelled and the transect width.

Bird behaviour was recorded using the internationally accepted behavioural catalogue proposed by Camphuysen & Garthe (2004). This scheme was used for all ship-based surveys during the TEST-BIRD project. Flight heights of different seabird species were also estimated or measured. Flight heights were measured using the Vector 21 AERO Rangefinder (Vectronix AG). The rangefinder is a set of binoculars that can additionally measure distances and flight heights using a laser. At least two measurements per bird were made with this rangefinder. Only if these measurements were very similar, was the data used for further analysis.

As part of the TESTBIRD project, eight multiple-day ship-based surveys, and twenty-one aerial surveys, were carried out in the key study area and also in the reference area. Data was also used from all EIA studies in and around the key study area.

11.2.2 Data analysis

Distribution patterns

Construction of the *alpha ventus* windfarm started in September 2008 with the installation of the transformer platform. The twelve wind turbines were erected between April and November 2009. Thus, all data from EIAs conducted prior to September 2008 was used for the baseline analysis. Data for the period 2010 to 2012, both from all EIAs conducted near *alpha ventus* and from the TESTBIRD project, was used to illustrate the distribution of selected seabird species in relation to *alpha ventus*. A total of six species or species groups were analysed and their distribution during their most important period(s) of the year illustrated for the pre- and post-construction periods in the key study area. Data was collated over relatively large periods, often two seasons (Garthe et al. 2007), to make use of much better survey coverage. Abundances of birds were collated in grid cells of 1 × 1 km. They are given as individuals per km² and colour-coded on a species-specific basis by abundance. Data was not further interpolated to allow for spatially explicit data presentations. Only data with good observation conditions (ship-based surveys: seastate < 5 Bft) was used and corrected with a species-specific correction factor (Markones & Garthe 2010).

Changes of abundance

The same data as stated in the previous section was used to study potential effects of the operation of *alpha ventus* on the abundance of two species. Us-

ing modern statistical methods, the abundances of the two most numerous species, lesser black-backed gulls and common guillemots, were compared for the pre- and post-construction period. The variance of abundance over single survey days was compared for different distances from the windfarm. For this purpose, generalised linear mixed models (Faraway 2006) using the *glmer* function (*lme4* package, Bates et al. 2011) were applied in R 3.0 statistics software (R development core team 2012). The number of individuals counted within each unique distance from *alpha ventus* (summarised in the following ranges: 0–2 km, 2–6 km and 6–10 km) was consequently tested in three separate models for significant differences using a Poisson error distribution.

Behavioural observations

Sample sizes for behavioural observations of lesser black-backed gulls were large enough to allow for detailed analyses. This was done using two four-to-five day surveys in May 2010 and June 2011. The following behavioural categories were used: Searching for food, actively feeding, and resting (including preening and bathing). All other behaviours were categorised as 'other'. In the event that more than one behavioural category was observed per individual, only one behaviour was used: Either the one not affected by the counting ship (e.g. if a bird was flushed by the ship movement) or the one considered more important (e.g. feeding actively over merely searching for food). The percentage for each behavioural category was calculated separately for the key study area and for the reference area.

Flight heights

To analyse flight heights of seabirds, measurements were taken during ship-based surveys conducted as part of the TESTBIRD project. For this analysis, flight heights of six seabird species that were measured using the rangefinder in 2010 and 2011 are shown in a boxplot and set in relation to the operational height of the rotor blades. Estimates of flight heights made only relative to the sea surface or the windfarm were not considered. All measurements analysed were from individuals that were neither associated with *alpha ventus* nor with the observation platform and any other ship nor aircraft and

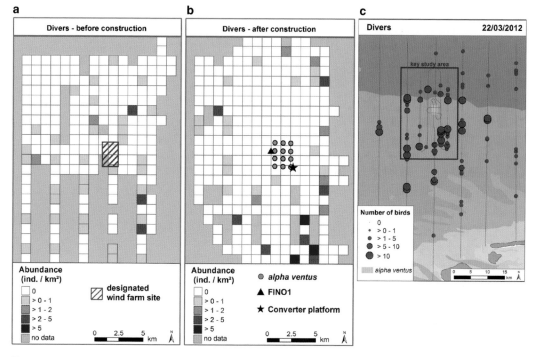

Fig. 11.4 Distribution of divers (red-throated and black-throated divers combined) before (a) and after (b) construction of *alpha ventus*. Data originates from ship-based surveys conducted from March to mid-May in the years 2000–2008 and 2010–2012, respectively. Additionally, (c) shows the results of one aerial survey with high numbers of divers conducted on 22 March 2012. Black lines illustrate the survey transects, red dots show the occurrence of divers.

can thus be considered undisturbed. Flight heights measured in both the key study area and the reference area were therefore pooled.

11.3 Results

11.3.1 Distribution patterns

In most of the species investigated in detail, maps show differences in distribution between the pre- and post-construction periods at *alpha ventus*. In four of the six species (groups), overall abundance was apparently lower after the windfarm was completed.

Red-throated and black-throated divers: Divers were relatively scarce in the key study area both before and after construction of *alpha ventus*. While there were three sightings with six individuals within the designated area of the windfarm before construction (Fig. 11.4a), no diver was seen afterwards – either during ship-based surveys

(Fig. 11.4b) or during aerial surveys (not shown). The closest sighting from the outer turbines was 1.1 km. Although the species are not very abundant in the key study area, they appear to actively avoid the windfarm; this is most apparent in the aerial survey, with the highest number of sightings per day occurring during the TESTBIRD surveys as illustrated in Fig. 11.4c.

Northern gannet (Fig. 11.5): Northern gannets' overall abundance was generally very low in the key study area and even lower after construction. While individuals were seen seven times (nine individuals in total) within the designated windfarm area before construction, none have been observed entering the windfarm area since construction. The closest sighting to the windfarm was 1 km.

Little gull (Fig. 11.6): Overall abundance of little gulls was considerably higher after construction of *alpha ventus*. This held true for most of the key study area, while the highest abundances after construction were at distances of 3 to 10 km from the outer turbines.

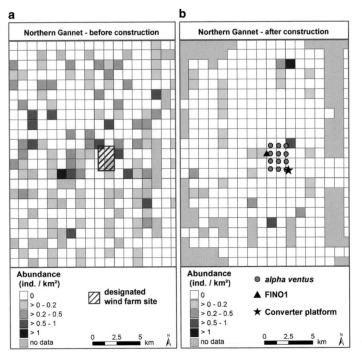

Fig. 11.5 Distribution of northern gannets before (a) and after (b) construction of *alpha ventus*. Data originates from ship-based surveys conducted from March to September 2000–2008 and 2010–2012, respectively.

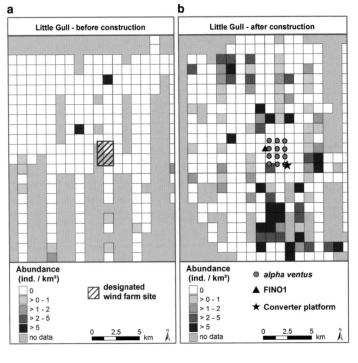

Fig. 11.6 Distribution of little gulls before (a) and after (b) construction of the *alpha ventus* windfarm. Data originates from ship-based surveys conducted from April to May 2000–2008 and 2010–2012, respectively.

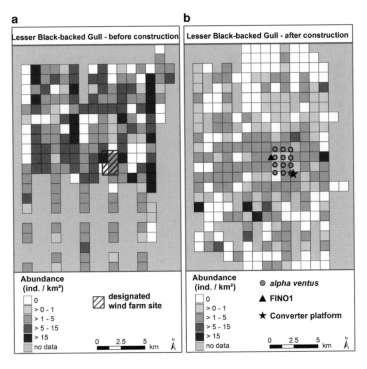

Fig. 11.7 Distribution of lesser black-backed gulls before (a) and after (b) construction of the *alpha ventus* windfarm. Data originates from ship-based surveys conducted from May to July 2000–2008 and 2010–2012, respectively.

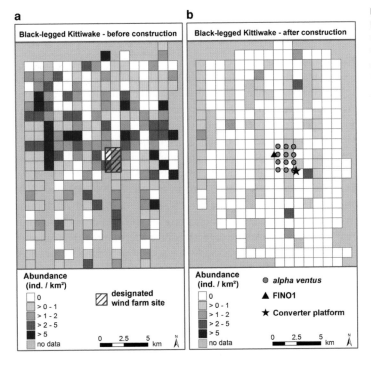

Fig. 11.8 Distribution of black-legged kittiwakes before (a) and after (b) the construction of the *alpha ventus* windfarm. Data originates from ship-based surveys conducted from November to April 2000–2008 and 2010–2012, respectively.

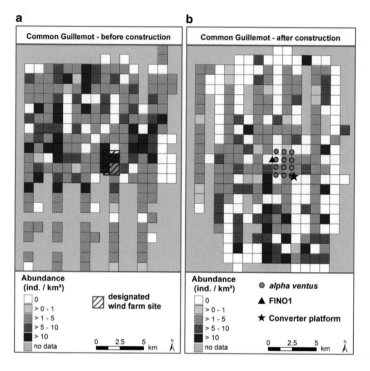

a

b

Fig. 11.9 Distribution of common guillemots before (a) and after (b) construction of the *alpha ventus* windfarm. Data originates from ship-based surveys conducted from November to February 2000–2008 and 2010–2012, respectively.

Lesser black-backed gull (■ Fig. 11.7): This species occurred in the highest abundance of all species during the breeding season (May to July). A clear decrease in overall numbers becomes apparent when comparing abundance values for the pre- and post-construction period. Lesser black-backed gulls occurred regularly inside *alpha ventus* after construction, but abundances were low to medium there. The highest abundances were found at a few kilometres distance from the windfarm site.

Black-legged kittiwake (■ Fig. 11.8): Although a few individuals were sighted inside *alpha ventus* during the operational period, black-legged kittiwakes showed a remarkable decline in overall abundance after construction. The decrease occurred throughout the study area.

Common guillemot (■ Fig. 11.9): Winter distribution of common guillemots changed from higher abundances in the northern part of the study area before construction of *alpha ventus* to higher abundances in the southern part after construction. Average abundances were lower after construction. Guillemots were seen inside the windfarm, but numbers were relatively low.

11.3.2 Changes in the abundance of lesser black-backed gulls and common guillemots

Significantly lower abundances of both lesser black-backed gulls and common guillemots were observed after construction of *alpha ventus* compared to before. This phenomenon occurred at different distances from the windfarm. The number of common guillemots was significantly lower in all distance classes ($0 - 2$ km: $b = -1.41$, $z = -3.89$, $p < 0.001$; $2 - 6$ km: $b = -0.88$, $z = -2.97$, $p = 0.003$; $6 - 10$ km: $b = -0.88$ $z = -3.20$, $p = 0.001$), indicating that the presence and operation of the wind turbines had a negative effect on the abundance of this species. As suggested by the model, this effect was highest within 2 km of the windfarm area. The number of lesser black-backed gulls decreased significantly within the 2 km distance class ($b = -1.38$, $z = -2.02$, $p = 0.043$), within the $2 - 6$ km distance class ($b = -1.07$, $z = -2.12$, $p = 0.034$), and also within the $6 - 10$ km distance class ($b = -0.88$, $z = -3.20$, $p = 0.001$). Based on the predicted values, the disturbance effect was strongest within 2 km of the windfarm.

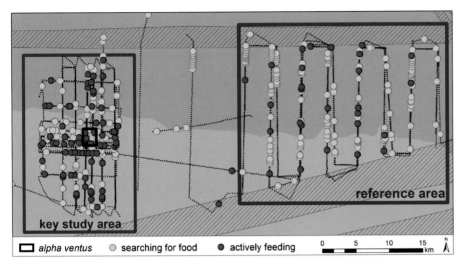

Fig. 11.10 Spatial distribution of lesser black-backed gulls searching for food (yellow dots) and feeding actively (red dots). The number of individuals per position (dot) is not indicated. Black lines show the routes sailed by the ship from which the observations were made.

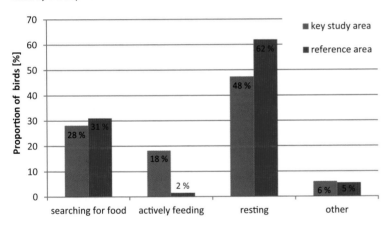

Fig. 11.11 Behaviour of lesser black-backed gulls in the key study area and in the reference area. The data base comprises all individuals for which a behaviour could be determined.

11.3.3 Behavioural observations

Overall, there were marked differences between the abundance of lesser black-backed gulls in the key study area and the reference area. In the key study area, 1.6 ind./km on effort were seen of which a behaviour could be allocated to 0.7 ind./km. In the reference area, overall abundance comprised 6.8 ind./km on effort, with 3.0 ind./km with behavioural recordings.

Lesser black-backed gulls showed activities involving finding and consuming food ('actively feeding') throughout the entire study area (Figs. 11.10, 11.11). 'Searching for food' was almost equally distributed over both surveyed study areas.

Feeding occurred in high frequencies outside *alpha ventus*, yet in close vicinity of the windfarm. In contrast, resting activities were more dominant in the reference area than within range of the windfarm (Fig. 11.11).

Taking a more detailed look at the findings, from a total of 219 individuals 'searching for food' within the key study area, five lesser black-backed gulls were associated with fishing vessels, while seven occurred within the windfarm. Some 62 out of 142 'actively feeding' birds were seen with the windfarm but none was associated with a fishing vessel. From a total of 368 'resting' birds in the key study area, seven were seen resting within the windfarm. In the reference area, commercial

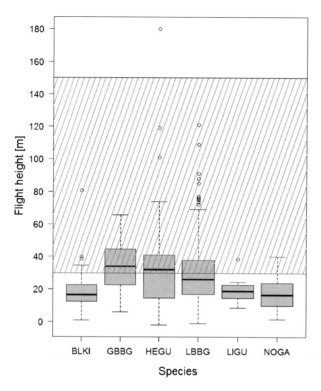

□ Fig. 11.12 Comparison of flight heights measured with the rangefinder for selected seabird species: BLKI (black-legged kittiwake, *n* = 36), GBBG (great black-backed gull, *n* = 25), HEGU (herring gull, *n* = 25), LBBG (lesser black-backed gull, *n* = 637), LIGU (little gull, *n* = 17) and NOGA (northern gannet, *n* = 24). One outlier (287 m) in the flight height of a lesser black-backed gull is not shown in the boxplot. The boxes show 25–75 % of all values, the bold black line indicates the median (50 %). Black dotted lines indicate the 95 % confidence intervals, and circles are outliers in the data sets. The red-hatched box illustrates the operational height of the rotor blades. Please note that all measurements shown here were carried out outside *alpha ventus* and were thus independent of the turbines.

fishing activities are much more pronounced: 730 out of 929 'food-searching', as well as 1,350 out of 1,856 'resting' lesser black-backed gulls were associated with fishing vessels. However, only two out of 46 'actively feeding' birds were associated with them.

11.3.4 Flight heights

Analyses of flight heights show that black-legged kittiwakes, little gulls and northern gannets prefer flight heights between 10 to 20 m (□ Fig. 11.12). Median flight heights of these species were between 15 to 18 m. Only a few measurements reveal potential overlaps with the rotor swept area. The three larger gull species, great black-backed gull (*Larus marinus*), lesser black-backed gull and herring gull flew considerably higher on average. Their flight heights could theoretically overlap with the rotor swept area. Median flight heights of great black-backed gulls and herring gulls were between 30 to 35 m, although the sample size for these two species is much smaller than that for

the lesser black-backed gull. The lesser-black-backed gull was by far the most abundant species for which flight heights were measured. The findings for this species can thus be considered very reliable. Less than half of the individual measurements comprised heights above 30 m, with the consequence that many lesser black-backed gulls are exposed to a possible collision risk with the rotor blades.

11.4 Discussion

Although plans to construct offshore windfarms exist since 2000 and several projects have been approved in the North Sea, experience from operating windfarms is still far from being complete, and especially for the German North Sea many aspects remain unclear. In particular, assessment of possible long-term effects is difficult due to a lack of long-term studies on windfarms. Such information would be extremely important for the planned further expansion of offshore wind energy in the North Sea and other sea areas.

Generally, seabirds can be affected by offshore windfarms in four different ways (e.g. Dierschke & Garthe 2006):

(1) Collision: Birds may collide with the wind turbines and are likely to be killed in such situations.

(2) Barrier effect: Birds may avoid the windfarm by flying around (horizontal escape) or over it (vertical escape). Such escape movements involve additional energy costs for the birds due to the detour and additional flight activities.

(3) Habitat loss: Resting and foraging seabirds may avoid windfarm areas because of the disturbance from the technical structures.

(4) Attraction: Birds may be attracted by the 'irregularity' of the sea surface, and also by potentially increased food availability due to a ban on fishing activities.

A review of the literature on the effects of operating windfarms found that several seabird species wholly or partly avoided windfarm areas. This was found, for example, in red-throated divers and included not only the windfarm area itself, but also a buffer zone of at least 2 km around the site (Dierschke & Garthe 2006). It is thus reasonable to expect that such birds as they occur at proposed windfarm areas will lose their habitats after completion of the windfarms.

No such data has been available for German waters to date. The investigations at the *alpha ventus* test site provide new and important insights into these effects. From the current state of analysis we can conclude that the two most numerous species of the study area, lesser black-backed gull and common guillemot, occur in significantly lower numbers after construction of *alpha ventus* than before. The same becomes apparent for other, less numerous species such as black-legged kittiwake and northern gannet while overall diver numbers may not have decreased much. Little gulls occurred much more numerous during the spring surveys in the post-construction period. The distribution maps (◘ Figs. 11.4 to 11.9) for the post-construction period clearly show that the highest abundances of all species always occurred outside the windfarm, usually at several kilometres distance. This was independent of whether a species has increased or decreased in number overall. While decreasing or increasing trends of less numerous and/or migrating species such as northern gannet and little gull may be a result of the occurrence at a much larger scale than analysed, the complete or partial avoidance of *alpha ventus* and nearby waters (a few kilometres depending on the species) by several species appears to be a stable pattern.

Possible effects of windfarms may not only be visible in changing abundances and/or distribution patterns. Species may also respond by altering their behaviour. Thus, the behaviour (e.g. searching for food, resting) and flight heights of all seabird species were observed inside and outside *alpha ventus*. Although the overall abundance of lesser black-backed gulls was much lower in the key study area than in the reference area, the proportion of birds searching for food was relatively similar. Both areas thus appear suitable for foraging. However, actively feeding birds were observed proportionally more often in the key study area. Almost half of these lesser black-backed gulls even fed within or very close to *alpha ventus*. This might be due to new hard substrate or small scale turbulence around the wind turbines that could provide an increased food supply. In the reference area, only few actively feeding gulls were seen, while numbers of resting birds were high. Many of these birds were associated with fishing vessels and were probably digesting or waiting for discards to be discharged. For most other species, sample sizes on behaviour obtained were too low to allow for any conclusions on the functionality of the key study area after construction of *alpha ventus*.

Flight heights were successfully measured in various species outside the windfarm. These measurements may be used as basic information for preferred flight heights of undisturbed birds. The measurements indicate potential small to large overlaps in flight heights of seabird species with the rotor swept area of the wind turbines. Overlaps were strongest in gulls (i. e. in species that were often recorded inside *alpha ventus* in the post-construction period), suggesting a higher risk of collisions in these species. Such information may become very relevant when comparing flight heights inside and outside windfarms (our sample size is currently still too low) and as input for collision risk models.

11.5 Perspectives

At this stage, it remains unclear to what extent results from a small windfarm, such as the *alpha ventus* test site, can explain the effects of much larger windfarms. It should further be considered that not all of the studied seabird species have their main distribution in the study area. The extent of windfarm-related effects is therefore not predictable for other designated project areas with major seabird concentrations. What was surprising, however, was the evidence of significant effects on numbers even in species that were expected to respond little to offshore windfarms (e.g. lesser black-backed gull) indicating that effects may also be substantial for these species when installing windfarms over large areas. To assess such effects, studies are needed at the first large-scale windfarms in the German Bight. Also, the issue of potential habituation cannot be answered at present. In one prominent example from an operational windfarm in Danish waters (Horns Rev 1), no habituation was observed in divers, even after five or six years. These species still avoid the area (Petersen & Fox 2007, Petersen et al. 2008).

Also, the effects of ship and helicopter traffic during operation must be studied more intensively, as species such as divers and sea ducks respond negatively to ship traffic (Schwemmer et al. 2011). Again, the intensity of routine traffic caused by windfarms will only be apparent a few years after construction, thus underlining again the importance of ongoing studies to investigate possible windfarm effects.

11.6 Acknowledgements

We are grateful to various people and institutions who supported us throughout the various stages of the two projects: Numerous observers during the ship-based and aerial surveys; M. Boethling, A. Beiersdorf, K. Blasche and A. Binder from the Federal Maritime and Hydrographic Agency (BSH); T. Verfuß from Project Management Jülich (PtJ); the consulting companies for their cooperation; the crews of the MVs Christoffer and Laura; and the crews of the planes from Sylt Air and BioFlight A/S; and A. Webb, R. Hexter and M. Robinson from HiDef Ltd.

Information box: High definition digital imagery aerial survey techniques

Visual survey techniques to estimate abundances of seabirds and marine mammals have traditionally been performed from ships or aircraft. From these platforms, seabirds and marine mammals are counted directly by the observers. Depending on the species and the research focus, either ship-based or aerial surveys may be better suited (e.g. Garthe et al. 2004, Camphuysen et al. 2004). Aerial surveys for seabirds are conducted traditionally at flight altitudes of 250 ft (76 m), and for marine mammals at 600 ft (183 m).

Recently, along with the construction of offshore windfarms, new safety concerns regarding low-flying aircraft in/near operating windfarms and/or windfarm construction areas became an issue. The development of high definition digital imagery survey techniques overcame this problem (◘ Fig. 11.13). Using these relatively new techniques, it will be possible to fly sufficiently high above operational wind turbines and wind-

farm construction sites. Further advantages of these techniques are that birds are not disturbed and flushed by low flying aircraft. With that, they prevent possible double counting of those birds. Additionally, numbers of birds can be accurately recorded and revisited, and a permanent observation record can be obtained.

Currently, there are two different methods that use high definition digital imagery in aerial surveys. These methods use either still cameras or video cameras. Both systems apply the very latest in camera techniques using high resolution cameras. Flight height with both techniques is usually between 1,476 ft (450 m) and 3,281 ft (1,000 m), and is thus located well above operational wind turbines or construction sites. Pictures and videos are recorded in flight, while the recorded data are analysed post-flight. Both techniques are coupled with a global positioning system (GPS) that connects each picture or video

◘ **Fig. 11.13** One of the survey planes used during aerial digital seabird surveys at *alpha ventus* (photo: HiDef Aerial Surveying Ltd.). View from the aircraft camera detecting assorted gulls and razorbills (photo: Ib Krag Petersen / Aarhus University).

frame with a geographical position. This enables the statistical analysis of the spatial distribution of seabirds and marine mammals. Still image and video cameras are programmed in advance to allow for automated continuous recording of transect strips. Ground resolution in all these systems ranges from 1 to 3 cm, enabling identification of animals – often to species level or at least species groups.

New software is being developed with a view to reducing the time needed to analyse the data post-flight. This new software should automatically extract possible objects from the images and help identify species.

Literature

Bates D, Maechler M, Bolker B (2011). Package "lme4" for the R software: Linear mixed-effects models using S4 classes.

Camphuysen CJ & Garthe S (2004). Recording foraging seabirds at sea: standardised recording and coding of foraging behaviour and multi-species foraging associations. Atlantic Seabirds 6: 1–32.

Diederichs A, Nehls G, Petersen IK (2002). Flugzeugzählungen zur großflächigen Erfassung von Seevögeln und marinen Säugern als Grundlage für Umweltverträglichkeitsstudien im Offshorebereich. Seevögel 23: 38–46.

Dierschke V & Garthe S (2006). Literature review of offshore windfarms with regards to seabirds. In: Zuccho C, Wende W, Merck T, Köchling I, Köppel J (eds): Ecological research on offshore windfarms: international exchange of experience. Part B: Literature review of ecological impacts: pp. 131–198. BfN-Skripten 186.

Faraway JJ (2006). Extending the Linear Model with R: Generalized Linear Mixed Effects and Nonparametric Regression Models. Boca Raton, Chapman and Hall.

Garthe S, Hüppop O, Weichler T (2002). Anleitung zur Erfassung von Seevögeln auf See von Schiffen. Seevögel 23: 47–55.

Garthe S, Dierschke V, Weichler T, Schwemmer P (2004). Rastvogelvorkommen und Offshore-Windkraftnutzung: Analyse des Konfliktpotenzials für die deutsche Nord- und Ostsee. Abschlussbericht des Teilprojektes 5 im Rahmen des Verbundvorhabens "Marine Warmblüter in Nord- und Ostsee: Grundlagen zur Bewertung von Windkraftanlagen im Offshorebereich (MINOS)

Garthe S, Sonntag N, Schwemmer P, Dierschke V (2007). Estimation of seabird numbers in the German North Sea throughout the annual cycle and their biogeographic importance. Vogelwelt 128: 163–178.

Markones N & Garthe S (2010). Ermittlung von artspezifischen Korrekturfaktoren für fluggestützte Seevogelerfassungen als Grundlage für Bestandsberechnungen von Seevögeln im Rahmen des Monitorings in der deutschen Ausschließlichen Wirtschaftszone von Nord- und Ostsee. Endbericht für das Bundesamt für Naturschutz.

Mendel B, Sonntag N, Wahl J, Schwemmer P, Dries H, Guse N, Müller S, Garthe S (2008). Profiles of seabirds and waterbirds of the German North and Baltic Seas. Distribution, ecology and sensitivities to human activities within the marine environment. Naturschutz und Biologische Vielfalt 59.

Noer H, Christensen TK, Clausager I, Petersen IK (2000). Effects on birds of an offshore wind park at Horns Rev: Environmental impact assessment. NERI Report 2000.

Petersen IK & Fox AD (2007). Changes in bird habitat utilisation around the Horns Rev 1 offshore windfarm, with particular emphasis on Common Scoter. NERI Report, commissioned by Vattenfall A/S, DK.

Petersen IK, Fox AD, Kahlert J (2008). Waterbird distribution in and around the Nysted offshore windfarm, 2007. NERI Report, commissioned by DONG Energy, DK.

Pihl S & Frikke J (1992). Counting birds from aeroplane. In: Komdeur J, Bertelsen J, Cracknell G (eds): Manual for Aeroplane and Ship Survey of Waterfowl and Seabirds. IWRB Special Publication 19: 8–23.

R Development Core Team (2012). R: a language and environment for statistical computing. R Foundation for Statistical Computing, Vienna, Austria.

Schwemmer P, Mendel B, Sonntag N, Dierschke V, Garthe S (2011). Effects of ship traffic on seabirds in offshore waters: implications for marine conservation and spatial planning. Ecological Applications 21: 1851–1860.

Skov H, Durinck J, Leopold MF, Tasker ML (1995). Important Bird Areas for seabirds in the North Sea. BirdLife International, Cambridge.

Tasker ML, Hope Jones P, Dixon T, Blake BF (1984). Counting seabirds at sea from ships: a review of methods employed and a suggestion for a standardized approach. Auk 101: 567–577.

11

Of birds, blades and barriers: Detecting and analysing mass migration events at *alpha ventus*

Reinhold Hill, Katrin Hill, Ralf Aumüller, Axel Schulz,
Tobias Dittmann, Christoph Kulemeyer, Timothy Coppack

Federal Maritime and Hydrographic Agency,
Federal Ministry for the Environment, Nature Conservation and Nuclear Safety (Eds.)
Ecological Research at the Offshore Windfarm alpha ventus,
DOI 10.1007/978-3-658-02462-8_12, © Springer Fachmedien Wiesbaden 2014

12.1　Introduction

Given the ambitious plans to make extensive use of offshore wind resources in Europe, with wind-farm installations covering hundreds of square kilometres in both the North Sea and the Baltic Sea, there is growing awareness that wind turbines, which protrude into the open airspace, affect birds flying at sea (Drewitt & Langston 2006). The seasonal phenomenon of bird migration itself shows tremendous variation both from day to day and from year to year. However, risk assessment studies of migratory birds in relation to planned offshore windfarms have been routinely based on non-continuous observations from ships. This allows access to any offshore planning area but restricts sample sizes and the explanatory power of the data. To close this methodological gap, the German government has funded research projects such as the StUKplus programme leading to the installation of automated radar and camera techniques that enabled continuous monitoring of birds at the *alpha ventus* windfarm. Drawing on long-term data collected with these remote sensing devices from the FINO1 research platform, from the transformer station and directly from one of the wind turbines, we are beginning to gain a detailed picture of bird migration during the post-construction phase at *alpha ventus* (since 2010). The unique dataset, which includes baseline information since 2003, allows us to analyse the external factors that trigger behavioural changes in response to operational wind turbines, which in turn influence the risk of collision for individual birds.

12.2　Birds, blades and barriers

The sheer abundance and variety of birds combined with their unrivalled ability to fly and migrate means that there is hardly a spot on the globe where no bird is found. This means that wherever a wind turbine is installed a flying bird may get injured or killed by colliding with the rotor blades merely by chance. While there is ample evidence from carcass searches under onshore wind turbines, there is so far no way to directly count fatalities at offshore windfarms. Collision probabilities therefore have

to be inferred from the frequency of birds recorded in close proximity to wind turbines.

The risk of collision with vertical structures is expected to be high in environments that lack anticipatory visual cues (Martin 2011). Thus birds crossing the open sea may be particularly prone to collisions, especially at night when visibility is low. The incidence of collisions may also vary considerably among bird species, depending on sensory and behavioural capacities, and also on ecological context. Stochastic models predicting the population consequences of additional mortality are therefore limited and need to be complemented by measures of actual flight behaviour and collision risk under a range of situations. Besides birds hitting rotor blades directly, windfarms may also act more generally as barriers to movement, inducing behavioural changes that could increase the energetic costs of flight and reduce the probability of survival.

Most ornithological studies in the offshore wind sector have focused on **day-time movements** of waterfowl and seabirds and are mainly based on visual observations under fair weather, often in combination with radar measurements. Day-time migrants or foraging water birds flying between breeding, staging and feeding sites have often been shown to avoid flying through windfarms (e. g. Desholm & Kahlert 2005), resulting in longer flight routes that may entail higher energetic costs.

Little is known about the **nocturnal behaviour** of birds in relation to offshore windfarms. Eider ducks have been seen to show avoidance behaviour at night (at shorter distances from wind turbines than during the day) and to adjust their flight direction along rows of wind turbines (Christensen et al. 2004). Short-term avoidance movements by small birds have been observed at night near onshore windfarms (Winkelman 1992). For the majority of nocturnal migrants crossing the open sea (such as songbirds and waders), however, there is hardly any information relating to offshore windfarms, mainly due to methodological deficiencies. Yet information on nocturnal migrants, which make up approximately two thirds of all migrants, is essential in order to validate avoidance rate estimates based on day-time observations and the outcome of generalised collision calculations.

A special point needs to be made about the effect of artificial light on night-migrating birds in marine and coastal environments. Artificial lighting can attract birds, in particular during dark and misty nights, resulting in collisions – a phenomenon long known from coastal and offshore lighthouses and lightships on nights of poor visibility with low cloud or rain (reviewed in Newton 2008; ▶ Information box *Birds at the lights*). Songbirds and waders in particular tend to accumulate in impressive numbers during nights of mass migration. Illuminated wind turbines in combination with increased vessel activity during construction and maintenance can lead to higher levels of light pollution and consequently to higher collision risks for bird species that are attracted to light.

Birds at the lights

» "When the Galloper light ship had white light, great numbers of birds were attracted to its lanterns, but now that the light is red, bird-visitors are almost unknown."
William Eagle Clarke (1912)

The many victims found at lighthouses, lightships and illuminated offshore platforms show that migratory birds crossing the open sea face considerable risk of collision with man-made structures (for a historical review, see Vaughan 2009). Collision rates on lightships in the North and Baltic Sea are cited as having ranged from 100 to 200 collisions victims per year (Hansen 1954). Birds in their hundreds may also collide each year with other offshore structures such as measurement masts and oil rigs. Although the circumstances leading to collisions with such static structures cannot be directly transferred to offshore wind turbines, historical evidence shows that high collision rates can occur at illuminated marine structures (◘ Fig. 12.1) during nights of low visibility (fog, drizzle). The psycho-physiological mechanisms and functional significance of avian phototaxis (light attraction) are as yet unknown, and the role of light quality and intensity is a subject of ongoing research. Most evidence suggests as a rule of thumb that the less light, the better it is for birds.

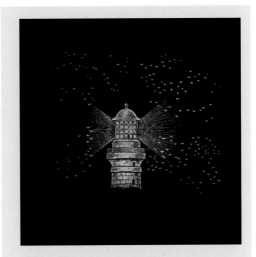

◘ **Fig. 12.1** Attraction of night-migrating birds by the lighthouse on Heligoland, North Sea, as depicted around 1895 by Heinrich Gätke on the cover of his seminal book, 'Heligoland as Ornithological Observatory'. At that time, avian collision victims were collected in sacks and used for human consumption.

12.3 Bird migration over the German Bight: Origin, phenology and species diversity

The North Sea is in the middle of the East Atlantic flyway, which is overflown twice a year by migratory birds. Millions of birds cross the German Bight on their recurrent journeys between breeding and wintering areas. An unknown proportion of birds cross the German Bight more than twice a year – for example irruptive migrants avoiding cold weather spells and foraging seabirds following stormy weather. 171 out of about 250 species registered each year in the southern German Bight have so far been detected in the vicinity of FINO1 over a period of ten years. Their breeding areas reach from north-eastern Canada in the west to Asian Siberia in the east and from Spitsbergen in the north to the Falkland Islands in the south. However, the majority of birds crossing the German Bight are from breeding areas in Scandinavia and a large fraction of these are songbirds, most species of which move to the south, southwest or west during autumn migration (Dierschke et al. 2011). Overall, bird migration over the German Bight takes place year-round with two clear migration peaks in

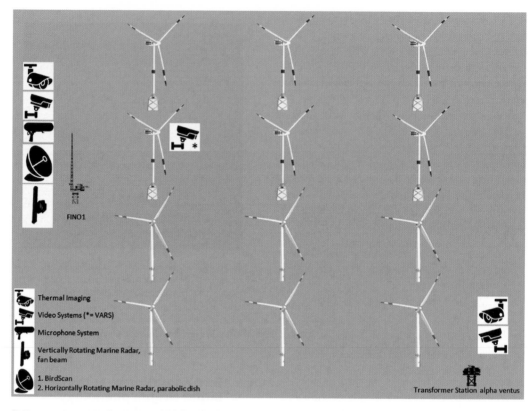

□ **Fig. 12.2** Overview of remote sensing techniques operating in the vicinity of *alpha ventus*.

spring and in autumn (Dierschke et al., 2011). Seasonal occurrence of birds is species-specific and migration intensity may differ strongly between spring and autumn as well as between species.

12.4 Methods

12.4.1 Methodological challenges

A key issue in bird migration research is how to increase the probability of detecting species at the individual level. Research is therefore driven by technological progress in remote sensing and bird marking techniques (Berthold 2001, Newton 2008). Any analysis of the potential effects of offshore windfarms on migrating bird species and populations thus depends on the availability of methods to increase detection probability. Various factors limit the study of detecting migrating birds in offshore environments:

(1) Species diversity: Bird species vary considerably in morphology, behaviour, abundance and ecology. With approximately 10,000 bird species worldwide and over 900 in the Western Palaearctic, it is hard to make generalisations from findings on single species.

(2) Nocturnality: About two-thirds of all migratory birds fly by night, significantly reducing the likelihood of their being sighted and counted by eye. Humans therefore depend on accessory sensors to be able to study nocturnal bird migration.

(3) Scale and dimension: The rotor-swept zone of an average offshore wind turbine is about twice the size of a football pitch. In contrast, most nocturnal migrants (songbirds) are no larger than golf balls. Detecting and counting birds in relation to the rotor-swept zone comes close to searching for needles in a haystack.

Ornithological research in the StUKplus programme set out to tackle these challenges by developing dedicated stationary radar and camera solutions for continuous monitoring of migratory birds at various spatial scales in and around *alpha ventus*.

To assess avian collision risks at offshore windfarms, birds need to be quantified within the rotor-swept zone and set in proportion to overall migration rates registered in and around the windfarm.

Various remote sensing devices (such as cameras and radars ▶ Information box *Remote sensing technologies*) have been installed in and around *alpha ventus*: On the FINO1 research platform in the west of the windfarm, on the transformer station in the southeast, and on one of the turbines (◼ Fig. 12.2).

Remote sensing technologies

Remote sensing is the acquisition of information about an object (such as a bird) or a phenomenon (such as bird migration) without physical contact with the object. There are two main categories of remote sensing methods: passive (e. g. photography and videography) and active (e. g. radar). All methods presented here were developed and/or installed in R&D projects funded by the German Federal Ministry for the Environment, Nature Conservation and Nuclear Safety (BMU).

Radar

Radar systems are a unique means of acquiring large amounts of bird migration data regardless of the time of day and across a range of weather conditions, although radar performance is affected by rain. Data quality deteriorates with increasing distance and decreasing bird size, resulting in a complex relationship between object distance and the probability of detection by radar (e. g. Eastwood 1967).

BirdScan

BirdScan is a purpose-built pencil-beam radar based on a conventional ship radar receiver and a parabolic antenna derived from the Swiss 'Superfledermaus' military tracking radar (Bruderer 1997, Neumann et al. 2009). It automatically detects and quantifies birds that fly through the distinct fixed radar beam. BirdScan's range is approximately three times larger than conventional ship radars set to the same power output (◼ Fig. 12.3). The sampling volume for calculating migration rates is well defined and the probability of detecting low flying birds is significantly higher than with conventional ship radars. In addition, BirdScan captures echo signatures (wing-beat patterns, ◼ Fig. 12.4) that allow clear separation of avian and non-avian radar signals as well as classification into various taxonomic groups (Zaugg et al. 2008). The results can be expressed as a migration traffic rate (MTR), defined as the number of bird echoes crossing a fictive horizontal

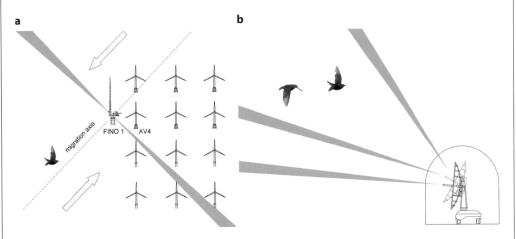

◼ **Fig. 12.3** (a) Schematic representation showing the beams of the dedicated BirdScan radar covering sections inside and outside of *alpha ventus*. An automated camera system (VARS, see below) was installed on wind turbine AV4. (b) Schematic representation of BirdScan measurements at different operation angles to cover multiple altitude layers.

a

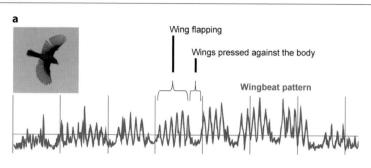

Wing flapping

Wings pressed against the body

Wingbeat pattern

b

■ **Fig. 12.4** (a) Wing-beat pattern of a songbird as recorded by (b) BirdScan radar.

line of one kilometre length per hour (Neumann et al. 2009).

Horizontally and vertically rotating marine radar

Horizontally operated marine radars are essential in determining migratory direction on the basis of flight patterns. Vertically oriented radars make for easy calculation of flight altitudes and intensities, although allowance must be made for bird detectability decreasing with distance from the device (Hüppop et al. 2004). It is nonetheless almost impossible to draw conclusions about species spectrum and flock size. The radars do, however, mostly capture the altitudes of birds out of reach of visual and acoustic registration and birds undetectable by large-scale weather radars due to the earth's curvature. Stationary marine radars are presently the only way of registering bird migra-

tion from just above the water surface to heights of more than 1.5 km. Available radar systems are not directly suitable for the investigation of collision risks because of their low resolution.

A self-rotating, parabolic dish antenna (Cooper et al. 1991) with a circular radar beam slightly inclined over the horizon is in operation at FINO1 (■ Fig. 12.5). This provides data on flight direction, including horizontal avoidance flights, without the interference from sea clutter experienced with earlier fan beam antennas. A complex algorithm in the radR software (Taylor et al. 2010) automatically combines single radar echoes over time into flight tracks. Eastwood (1967) showed in his seminal book about radar ornithology that a bird's radar cross-section and hence its detectability is directly affected by its aspect to the radar: Birds recorded from the front or back have just half the detection

● **Fig. 12.5** Marine radar with slightly upward-tilted *horizontally rotating parabolic dish antenna* at FINO1.

range of birds recorded laterally. Non-rotating radars always assume the same main flight direction axis, which does not hold true because of the varying wind conditions offshore. Rotating parabolic dish radar can account and compensate for this because the flight direction of each single bird is known. Marine radars with parabolic dish antennas have operated successfully since 2009 and 2010 at the FINO1 and FINO3 research platforms.

VARS

VARS (Visual Automated Recording System) is a camera system that automatically detects flying birds at day and night . The purpose-programmed motion analysis software saves the incoming video streams only when one or more objects move through the image. In darkness, infrared light (in an active system) enables the system to record birds and bats (Schulz et al. 2009) (● Fig. 12.6).

Automatic sound recording

Acoustic detection of bird calls has worked satisfactorily for many decades and has recently been standardised (Farnsworth 2005). Recordings at sea are often degraded by strong wind noise but the process can be automated with certain restrictions. It has to be noted that acoustic data collection is not suitable on its own for quantifying bird migration, as some bird species utter no calls during migration, while others increase call activity in poor visibility or when attracted to light. Nevertheless, the registration of bird calls is so far the only remote sensing technique that can provide reliable data on migration intensity at species level. Bird calls were detected and recorded close to FINO1 using a microphone with a windshield (● Fig. 12.7). The range of the microphone varies between species and weather conditions. Typical calls of thrushes like blackbird and redwing are detected up to 100 m roughly. The AROMA

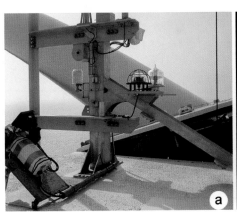
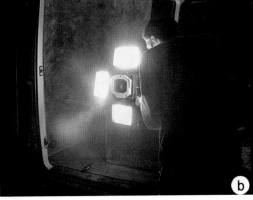

● **Fig. 12.6** (a) VARS on the nacelle of turbine AV4 at *alpha ventus*, (b) the infrared beams used at night are normally invisible to birds and humans (infrared-sensitive photo taken during nocturnal trial).

◘ Fig. 12.7 Microphone with wind shield at FINO1.

('Acoustic Recording of Migrating Aves') capturing software was developed to recognise bird calls automatically by their characteristic narrow sound spectrum while filtering out most wind and rain noise (Hill & Hüppop 2007, Hüppop & Hill 2007).

Automatic video and thermal imaging camera systems

Video cameras can be used for automated recording of daytime bird activity up to some hundred metres, depending on bird size. Using computers, images captured over a one-minute period are processed to create two separate peak-storage images, one image containing only the brightest pixels (peak), the other only the darkest (Hill & Hüppop 2007). This reveals flight tracks against both dark and light backgrounds, and also provides information on approximate directions and at least the species group and flock size. At night, a multi-megapixel video system with additional high-performance infrared illumination mounted on the transformer station was used to observe part of the rotor-swept zone of the closest wind turbine. Thermal imaging cameras were additionally operated on FINO1 and the transformer station (◘ Fig. 12.8). Thermal images show the shape of warm objects. The maximum range of some hundred meters depends on bird size and weather conditions. Fog and drizzly rain significantly reduce the range of vision of all cameras and the ability to identify species.

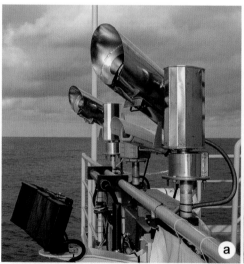

◘ Fig. 12.8 (a–b) Video cameras and thermal imaging devices at the *alpha ventus* transformer station observe part of the rotor-swept zone of a wind turbine.

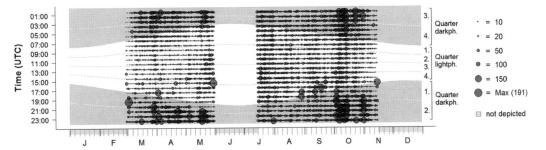

Fig. 12.9 Mean bird migration intensity (echoes/h; n = 859,957 echoes) at any hour at FINO1 during the main migration periods (2004–2012) measured at heights up to 1,000 m.

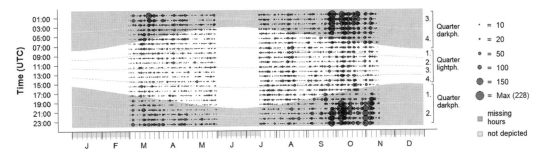

Fig. 12.10 Bird migration intensity (echoes/h; n = 36,580 echoes) at any hour UTC at FINO1 during the main migration periods in 2010 measured at heights up to 200 m.

12.5 Results and discussions

12.5.1 Measuring long-term patterns with radar and optical systems

A vertically rotating marine radar (vertical radar ▶ Information box *Remote sensing technologies*) is in use at FINO1 to reveal general patterns of offshore migration. In continuous operation since autumn 2003, the vertical radar represents the first time ever that bird migration intensity has been recorded nonstop day and night in an offshore area for a period of several years. This permits closer examination of potential changes in echo rates.

Data gathered up to a maximum height and distance of 1,500 m has sharpened the picture of birds crossing the German Bight and added invaluable insights in terms of risk assessment, thereby highlighting the phenomenon's complexity: Bird migration across the German Bight takes place throughout the whole year, but primarily in spring and autumn, and is usually more pronounced at night than during the daytime (**◼** Fig. 12.9, **◼** Fig. 12.10).

Temporal patterns vary between years, notably because very high migration intensities were recorded on only few nights or on a restricted number of days. Typically, both spring and autumn migrations occured in waves: A number of nights with strong migration were followed by a period with low intensity lasting several days (e. g. **◼** Fig. 12.10).

In all seasons, the highest intensities were measured in the lowest 200 m, meaning a large part of migration over the sea occurs at an altitude that would bring birds within the reach of wind turbines (**◼** Fig. 12.11). The radar system is well adapted to detecting bird flight altitudes at FINO1, even at a high-resolution time scale. Strong nightly migration often corresponds with high flight altitudes, both of which reflect favourable conditions for migration. Deteriorating conditions, however, easily force birds down and so expose them to collisions with offshore structures. This can happen within a few hours and is illustrated for a night in autumn 2010 (**◼** Fig. 12.12). The same night in fact proved to be the occasion of a mass fatality event at FINO1, which was well documented by various

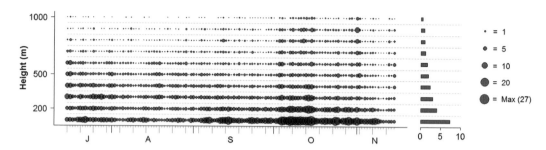

▣ Fig. 12.11 Nocturnal height profiles of migrating birds in autumn (2004–2012; no data available for 2007; *n* = 424,118 echoes) at FINO1 as measured by vertical rotating marine radar (left). An overall mean migration rate (echoes/h) is given per height class (right).

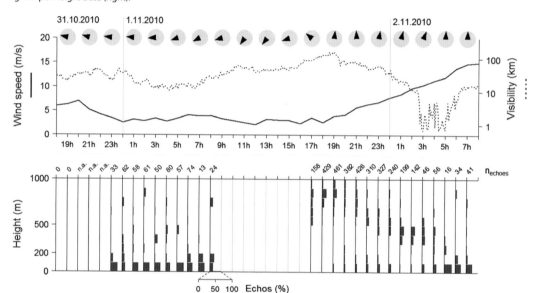

▣ Fig. 12.12 Migration intensities and nightly height profiles (without data during daylight) of migrating birds during two consecutive nights at FINO1 in autumn 2010 measured by vertically rotating marine radar (lower figure). Accompanying weather conditions (upper figure) are given as follows: wind direction (arrows), wind speed (solid line), visibility range (dotted line). A mass fatality event took place in the night of 1 to 2 November 2010.

remote sensing techniques and is described in detail below.

Vertical radar is a highly useful tool for measuring how many birds pass when and at what heights, and statistical methods permit causal analysis taking into account underlying weather conditions – a major goal of ongoing research effort. The ability to detect behavioural responses is limited, however, though there is huge additional demand for behavioural data to clarify collision mechanisms and for risk assessment.

A second marine radar has a horizontally rotating parabolic dish antenna (horizontal radar) to detect deviations from the expected main mi-

gration direction. This is well illustrated by the observed mass collision night of 1 to 2 November 2010 (▣ Fig. 12.13) when tracked birds flew in all directions, most of all close to the platform and the nearby turbines. There may also have been a certain amount of circling flight. As the night progressed and with upcoming foggy conditions, all operating radar systems (see below for a third radar type) consistently detected more and more birds. Whith the first daylight birds vanished.

It is revealing to compare the migration course during the collision night of 1 to 2 November 2010 with the situation at a similar research platform,

FINO3, located 136 km to the northeast of FINO1 and west of the island of Sylt: At FINO3, high but unexceptional migration intensity was recorded by an identical horizontal radar. Birds passing FINO3, however, still encountered fairly good weather conditions. They flew at higher altitudes and were headed towards the west and southwest (◘ Fig. 12.13), as expected for autumn migration. Such comparisons show that bird migration and constraining or supporting weather conditions can differ considerably between different sites within the German Bight at any one time.

Apart from providing such highly sought-after insights into behavioural responses relating to weather, horizontal radar also supplements the picture outlined by vertical radar: Because of its mounting position, the vertical radar detects birds only within a limited section outside of the windfarm. In contrast, horizontal radar scans nearly full circle around FINO1 every 2.5 seconds, thus allowing coverage of parts of the windfarm as well.

Differences consequently become visible between the figures illustrating the nights between 31 October and 2 November 2010. According to vertical radar, the night of 31 October to 1 November 2010 showed moderate migration intensity compared with the ensuing mass-collision night (◘ Fig. 12.12), whereas horizontal radar showed roughly equal intensities (◘ Fig. 12.14). This is in line with data obtained with the pencil beam radar, which confirmed especially high migration rates within the windfarm reflecting overall high migration intensity (see below).

To be able to tell the whole story, data was gathered continuously using several optical systems. Video and thermal imaging cameras at FINO1 recorded peak-storage images (▶ Information box *Remote sensing technologies*), providing a useful tool for analysing the behavioural reactions of specific species to a windfarm in daylight conditions. At night, a video camera aided by strong infrared illuminators and two thermal imaging cameras observed the rotor-swept zone of the turbine closest to the transformer station, albeit still from a distance of 250 m. Over three years of data acquisition, the two systems recorded hundreds of birds flying through or near the blades (◘ Fig. 12.15), though no direct collisions were observed. The weakness of optical

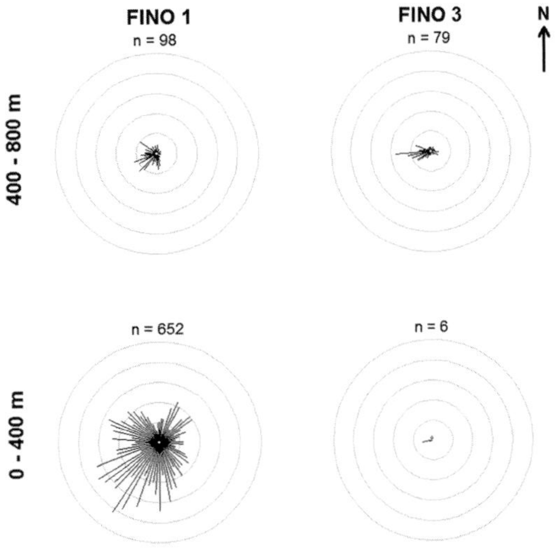

◘ Fig. 12.13 Direction of recorded bird tracks at FINO1 (left) and FINO3 (right) during the night of 1 to 2 November 2010 for two elevation levels (see vertical axis) measured by parabolic antenna. n-tracks are given in each case.

systems is their limited range in bad weather: With more water droplets in the air, visibility and hence the ability to detect birds is reduced. Nevertheless, all optical systems showed various bird signatures in the vicinity of *alpha ventus* during the night of 1 to 2 November 2010, but visibility was too low to observe birds directly in the rotor area.

Every detection method has its advantages and disadvantages. The lesson learnt is the need for thorough analysis comprising and combining different remote sensing techniques at various spatial scales.

12.5.2 'Ground-proofing' through automatic detection of species-specific bird calls

Many bird species utter calls during migration (e. g. Farnsworth 2005). Registered by a sensitive microphone and processed by specially developed software called AROMA (Automatic Recording of Migrating Aves), the system in use on FINO1 autonomously recognises bird calls by their characteristic narrow sound spectrum and filters out most wind, rain and wave noise. Bird calls are stored as audio files (Hill & Hüppop 2008) and recordings are subsequently matched to species by qualified staff. Given the re-

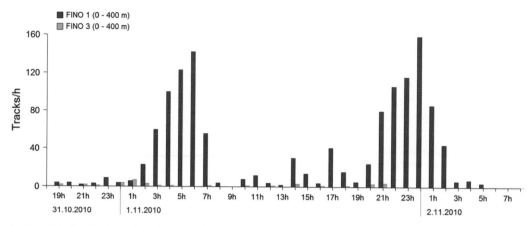

Fig. 12.14 Migration intensities of migrating birds during two consecutive nights in autumn 2010 at FINO1 and FINO3 as measured by parabolic antenna at lower elevations up to 400 m. Note that there was almost zero migration activity (at lower elevations only) at FINO3 and that intensities measured at FINO1 were roughly equal for the nights 31 October/1 November 2010.

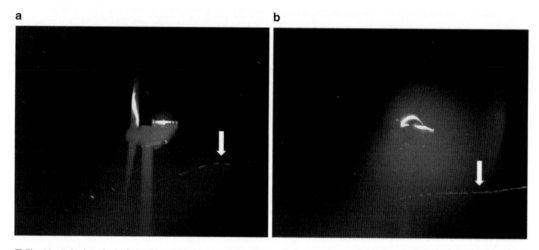

Fig. 12.15 Bird tracks (indicated by white arrows) passing through the rotating sphere of blades with different rotation axes as recorded by video camera. (a) Single images of a migrant bird form a track. (b) Parallel streaks show stars in the background.

stricted capture range of the microphone and the 112 bird species recorded, automated sound detection proves an excellent indicator for the presence of birds in direct vicinity of an offshore structure, showing that calling individuals are exposed to the risk of collision. It cannot be ruled out that individuals circling around the platform in poor visibility may have been recorded more than once. Our call rate data should therefore be thought of as a relative measure (which is sufficient in this context) rather than an absolute number of calling birds.

Call rates of nocturnally migrating thrushes at FINO1 rise when early-evening conditions are in-

dicative of favourable tailwinds. They are therefore likely to reflect 'upstream' take-off conditions (Hüppop & Hilgerloh 2012), as wind regimes, though highly variable, are similar over large areas. Interestingly, high call rates were also recorded under various adverse weather conditions such as favourable tailwinds turning into crosswinds/headwinds or reduced visibility. Thorough analysis, however, explained the apparent discrepancy: Hüppop & Hilgerloh (2012) conclude that unfavourable weather will cause high call rates only if preceding favourable winds (with supporting tailwind) and clear skies have induced mass migration in a main mi-

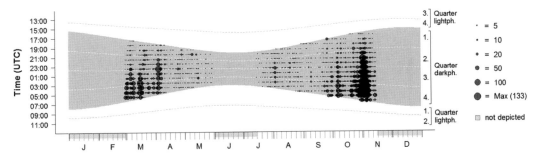

□ **Fig. 12.16** Mean nightly call rates (call-positive data files/h; $n = 70,704$) during the main migration seasons (2008–2012) measured by autonomous sound recording.

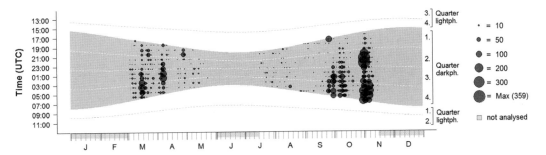

□ **Fig. 12.17** Nightly call rates (sums of call-positive data files/h; $n = 9,029$) during the main migration seasons in 2010 measured by autonomous sound recording.

gration season. The fact that no single factor ever causes call rates to drop to zero emphasises that factors should never be regarded separately. For example there may be many calling thrushes despite favourable low humidity because of winds becoming unfavourable. Hence extremely high call rates were registered only in comparatively few nights when certain parameters coincided.

Uncovering the environmental conditions of what actually leads night migrating birds into the sphere of illuminated offshore structures concurrently means unravelling the conditions that cause bird collisions. The physiological processes behind this behaviour, however, mostly remain in the dark.

As well as allowing causal analyses, the sound approach also makes it possible to pick out temporal patterns that are highly useful for risk assessment on both a seasonal and daily basis. □ Figure 12.16 illustrates the periods of time with relatively large numbers of events accompanied with high call rates. In spring, such periods occur mainly between late February and early April, while in autumn they are mainly concentrated

between the beginning of October and early November. When nocturnal patterns are taken into account, it becomes evident that high rates of bird calls predominate during the time after midnight. Apart from such general findings, however, bird call rates can peak at almost any time of year and any time of night. Focusing on single years, great variation becomes evident from night to night and even from hour to hour (□ Fig. 12.17). Such pronounced peaks suggest that high call rates can be attributed to singular events that can involve large numbers of birds as detected by radar; this is particularly evident at lower elevations (see above).

Overall, phenology data from sound recordings is roughly in line with the seasonal occurrence of collision victims found so far on the FINO1 platform, comprising 41 species and almost 1,000 individuals between autumn 2003 and end of 2012. The phenology data is biased towards the phenology of the genus *Turdus* regardless of whether the focus is on nightly rates of bird calls or collision victims: Thrushes of various kinds form the bulk of the data pool in both cases. Taking into account overall dif-

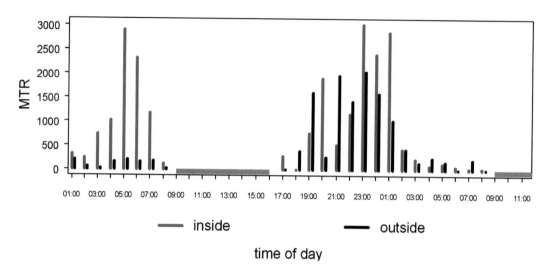

🔲 **Fig. 12.18** Migration traffic rate (MTR; bird echoes per hour and kilometre) of night-migrating birds measured with BirdScan radar inside (red) and outside (black) of *alpha ventus* on 1 to 2 November 2010 (grey horizontal bars: no measurement).

ferences in timing of migration between thrushes, stunning conformity emerges: Song thrushes *T. philomelos* turn up earliest in autumn and collide earliest, both as measured by median date. They are followed by redwings *T. iliacus*, which in turn are gradually superseded by blackbirds *T. merula* as autumn progresses.

As with high call rates, the lion's share of birds falling victim to collisions are attributable to specific events, with up to 200 birds found on a single occasion. Where collisions were traceable to a given night, they proved always to be accompanied both by a high frequency of birds at the lowest elevation levels (see below for case study) and by high bird call rates.

12.5.3 Detecting behavioural responses with fixed pencil beam radar

The aim of the ongoing survey with the BirdScan fixed beam radar (▶ Information box *Remote sensing technologies*) installed on FINO1 is to document and quantify the extent of behavioural changes in migratory birds (avoidance and attraction) in response to an operational offshore windfarm. BirdScan has been continuously measuring migration rates inside and outside *alpha ventus* since October 2010. The study design involves alternating measurements in-

side and outside the windfarm. The radar beam has a detection range of up to about 4,000 m in either direction. To detect migratory birds at different elevations, the parabolic antenna is set to three different angles on either side for distinct time periods. These evenly balanced, alternating measurements make it possible to detect spatiotemporal differences in bird numbers caused by behavioural responses (avoidance and/or light-induced attraction). Measurements were restricted to the night (i. e. to nocturnal migrants, which make up approximately two-thirds of all migrants), the explanatory power of radar surveys being limited during the day by the regular occurrence of bird flocks, which prevents individual echo classification. Most night-migrating birds, on the other hand, accomplish their journeys in solitary flights.

Investigating the spatial distribution of nocturnal songbird migration throughout a season, results with BirdScan show that within a migration season, total migration intensity can be distinctly higher inside the *alpha ventus* windfarm than outside. This was true most of all for low altitudes below 200 m. However, relative bird distributions can differ considerably from night to night (🔲 Fig. 12.18). Several mechanisms may be responsible for the specific bird distribution recorded by radar in a given night: Wind conditions, visibility and the operational status of the wind turbines. With respect to potential mitigation measures (e. g., shutting down turbines during peak migration), it is necessary to disentangle the potential factors leading to bird accumulations and collisions. As described below, one approach may be to compare consecutive nights that show strong bird migration but differ in weather, turbine activity, and the incidence of collisions.

12.5.4 Measuring bird movements through the rotor-swept zone with VARS

The VARS (Visual Automated Recording System; ▶ Information box *Remote sensing technologies*) camera system is mounted on the nacelle of the AV4 wind turbine. The camera automatically records day and night-time activity of birds moving through a sector of 22° parallel to the rotor blades. Given that, due to physical constraints, a compromise has to be found between the opening angle of the camera and distance-dependent image resolution, an opening angle of 22° was chosen to ensure a sufficient recognition of small passerines along the length of a rotor blade (about 60 m). Large birds are visible over much larger distances. The positioning of the camera allows the detection of birds even under low-visibility conditions (such as fog and drizzle). Changes in wind direction are compensated for as the integrated system moves with the nacelle. This significantly eases the extrapolation of measurements to the entire rotor-swept zone, which is not easily done with remote cameras that capture a constantly changing aspect of the rotor. With this approach, birds can be quantified inside the rotor-swept zone, ground-proofed, and set in relation to overall migration rates determined by radar. Combined with radar surveys, VARS allows an estimate of phototaxis and micro-avoidance. Bird collisions can then be treated as purely stochastic events on the basis of established collision models. Uncertainties associated with the use of avoidance rates in such models can be excluded through direct measurement of bird presence and performance within the rotor-swept zone. Estimates of avoidance rates on the basis of radar surveys or visual observations have previously been used to approximate the number of birds flying through the rotor-swept zone (measured here

directly with VARS). By comparing the frequency of birds measured in the rotor-swept zone with the extent of migration measured with radar (such as pencil beam radar) it has now become possible to quantify micro-avoidance on site.

In addition to the nacelle-mounted camera, a second VARS system was installed at the base of the tower on turbine AV4 to record bird activity in the area above the platform deck (20 m above sea level) in relation to turbine activity. Here, the distance between deck surface and rotor blade tips is only 9 m.

Detection of birds using VARS from the AV4 nacelle started in late September 2010, while measurements from the platform deck started in early March 2012. Besides the numerous birds flying high above the windfarm (mostly gulls), around 130 birds (approximately 50 % at day and night) were recorded within the rotor-swept zone through to the end of December 2012 (irrespective of turbine activity). Of all the recorded events, 91 % could be assigned to individual birds. Between March 2012 and December 2012, 5.6 times as many bird events were recorded on the platform deck than from the nacelle. The reason for this marked difference may be the greater attractiveness of the platform deck as a potential resting place for terrestrial species and the considerably stronger illumination of the deck and the tower base with white light. This also suggests that the probability of multiple observations of the same individual is much more likely on the platform deck than from the nacelle. Finally, avoidance of the rotor-swept zone may also contribute to the difference.

Birds (mostly gulls) were seen almost on a daily basis in the lower camera, with the highest activity levels measured from July to September. Some bird species were even registered in mid-winter (irruptive boreal migrants) and in summer (swifts). Thus, birds can be expected to occur near wind turbines throughout the year, emphasising the importance of continuous year-round measurement.

12.5.5 Species composition

Birds recorded with VARS can be assigned into broad species groups, while accurate classification

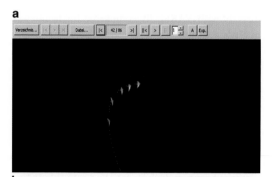

☐ **Fig. 12.19** (a) A swift *Apus apus* flying close to the stationary rotor of turbine AV4 at *alpha ventus* (digital still from a video sequence recorded with VARS in the night of 27 August 2011). (b) A wheatear *Oenanthe oenanthe* flying past the platform deck of turbine AV4 at *alpha ventus* (digital still from a video sequence recorded with VARS on 11 August 2012).

at species level is achieved only to a limited extent. However, the acquired images provide direct evidence for the range of potentially affected species. After excluding the omnipresent gulls from the analysis, songbirds dominated the species list (nacelle: 92 %, platform deck: 88 %). Starlings *Sturnus vulgaris*, thrushes *Turdidae*, crows *Corvidae*, larks *Alaudidae* and swallows *Hirundinidae* were the most frequently classified passerines. Among the seabirds, only one gannet *Morus bassanus* and one auk (razorbill *Alca torda* or guillemot *Uria aalge*) approaching the stationary turbines were detected by the lower camera. Once, several cormorants *Phalacrocorax carbo* landed on the platform deck, and a tern flew at deck height, safely avoiding the rotating turbine. Among raptors, a peregrine falcon *Falco peregrinus* and a kestrel *Falco tinnunculus* landed on the platform deck of the wind turbine, and a honey buzzard *Pernis apivorus* flew close past the inactive rotor. Swifts *Apus apus* were detected in the open airspace above the turbine on several days and nights in summer (☐ Fig. 12.19). Pigeons

Columbidae were seen regularly on the platform deck. Besides birds, both cameras also detected large numbers of insects. Despite clear evidence for large insects (butterflies *Lepidoptera*, flies, wasps, bees *Hymenoptera*, and dragonflies *Odonata*), problems in distinguishing birds and insects may arise at greater distances from the camera, especially at night when resolution decreases even further.

12.5.6 **Estimating collision risks in the rotor-swept zone**

As most collisions with birds are expected to occur in the rotor-swept zone, the operational state of the turbine needs to be taken into account when assessing net collision rates. Other collision risk factors to be considered include differences in turbine visibility at day and night and phototaxis in night-migrating birds under low-visibility conditions (▶ Information box *Birds at the lights*). We therefore analysed the bird events captured with VARS in relation to the operational state of the wind turbine, and separately for day and night.

Bird frequencies varied significantly between operational and non-operational phases. Over the entire study period, more bird events were registered from the nacelle when the rotor was standing still, both during the day and at night. However, this finding is purely correlational, and the precise cause-effect relationships between turbine activity, weather conditions, bird phenology and potential micro-avoidance need to be understood before drawing conclusion about the effectiveness of mitigation measures such as operational controls (shutting down wind turbines during peak migratory periods). Our findings with the VARS at AV4 suggest that the overall pattern of bird presence within the rotor-swept zone was driven by a few occasional events.

12.5.7 **A case of mass accumulation of night-migrating birds**

During a visit to the FINO1 platform on 5 November 2010, 88 carcasses of nocturnal migrants were found on the platform (Aumüller et al. 2011). Their injuries and body condition indicated that the birds

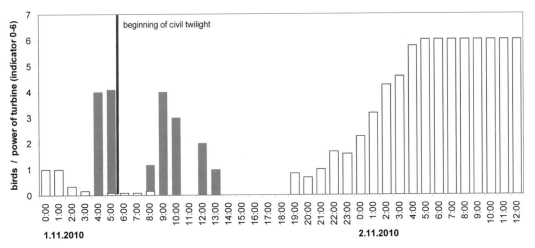

■ **Fig. 12.20** The number of birds recorded with VARS on 1 to 2 November 2010 in the rotor-swept zone of turbine AV4 (orange bars); white bars show turbine activity as relative score index, ranging from 0 (motionless) to 6 (maximum rotation rate).

had died after colliding with the platform and not of starvation. As the horizontal platform where carcasses could be retrieved (instead of falling into the sea) is comparatively small in relation to the vertical extent of the whole structure, the total number of actual collision victims was probably much higher. An unknown proportion of carcasses may also have been taken by gulls. From the preservation state of the carcasses together with video camera images and radar data from flying birds at sea on preceding nights, the night of 1 to 2 November 2010 was determined as the date of the collision event. This information could be combined with information from the VARS at the foot of the wind turbine to elucidate the factors leading to avian collisions with illuminated structures as well as the role of the rotating blades.

The collision night of 1 to 2 November 2010 was characterised by good migration conditions at the beginning of the night (light to medium tailwind and good visibility) with high bird flight altitudes. In the course of the night, migration conditions deteriorated with the wind changing to a headwind of increasing strength and visibility falling to about 700 m, as a result of which increasing numbers of migrating birds became concentrated in the lowest 200 m (Aumüller et al. 2011, ■ Fig. 12.12). Weather developments of this kind with increasing headwinds and decreasing visibility are typical for decreases in the flight altitudes of and phototaxis

in nocturnal migrants (Drost 1960, ▶ Information box *Birds at the lights*). Measurements with the BirdScan radar (▶ Information box *Remote sensing technologies*) confirmed strong bird migration at low altitudes for that night, with similar numbers of echoes recorded inside and outside the windfarm (■ Fig. 12.18). While many birds were evidently attracted to (and collided with) the illuminated research platform, no birds were visible at the time at the rotating wind turbine AV4 equipped with the VARS automatic camera system. In contrast, at the end of the previous night, several birds were recorded by VARS after the wind turbine stopped rotating (■ Fig. 12.20). Visibility and bird density in the windfarm were similar and weak tailwinds prevailed at both times. Interestingly, many more birds were recorded in that night by radar inside the windfarm than outside of it, indicating that the illuminated windfarm does attract nocturnal migrants.

These results confirm that illuminated windfarms can attract nocturnal migrants and influence the risk of collision, although their lighting differs from that of lighthouses and offshore platforms, for which numerous bird strike events have been documented in the past (Gauthreaux & Belser 2006). However, the results also show that bird strikes at FINO1 are not necessarily paralleled by bird strikes at a wind turbine nearby. In this case, the rotating rotor of the wind turbine surveyed with VARS might have reduced the occurrence of

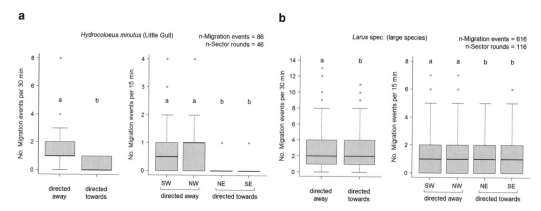

◘ **Fig. 12.21** Number of migration events per species (or species group) and viewing direction towards or away from *alpha ventus*, illustrating species-specific variety in response behaviour. Whereas little gulls (a) are much more frequent in sectors without turbines, large gulls (b) of the genus *Larus* occur regardless of viewing direction. Different letters (a, b) indicate significant differences.

birds in the rotor-swept zone, which is indicative of micro-avoidance. Nevertheless, it cannot be ruled out that birds were attracted to and collided with other wind turbines at *alpha ventus*, and/or that collisions with stationary turbine structures occurred in the night of 1 to 2 November 2010, because birds were also recorded by video at the turbine close to the transformer station, and collision risk can vary considerably between the turbines within the same windfarm (Langston & Pullan 2003). The results suggest that predictive collision models based on overall migration rates alone can lead to incorrect estimates. We conclude that migration rates need to be measured at multiple spatial scales and related to turbine activity in order to obtain more reliable estimations of avoidance and collision rates.

The brightly lit met mast of FINO1 with steel cables protruding some 100 m into open airspace probably contributed to the large numbers of collision victims found at the platform deck. At a platform of the same type in the Baltic Sea (FINO2), VARS recorded bird collisions with steel cables (mainly at dawn before sunrise) and with the steel grid construction (at night) (Schulz et al. 2011). At a third similar research platform (FINO3) west of the island of Sylt, just three dead birds were found on the deck in the same two nights, even though migration intensity there was similarly high. Most of the birds evidently passed FINO3 in better weather conditions and at higher altitudes than

FINO1. Nevertheless, the processes leading to collisions with these static structures are not necessarily transferable to wind turbines, which have different structural components (platform deck, shaft, nacelle and rotor blades) and different lighting conditions.

12.5.8 Windfarm-induced deviations in route under daylight conditions

At night, measurements to detect spatio-temporal differences in bird numbers caused by behavioural responses (avoidance and/or attraction) were restricted to the use of pencil-beam radar and VARS (see above). During daylight, visual observation at FINO1 additionally permitted behavioural observation and identification at species level within a much larger space (Aumüller et al. 2013). Bearing in mind that various bird species migrate strictly by daylight, and measurements of mortality mainly relate to darkness and illuminated offshore structures, there is a need for species-specific information on behavioural responses. Direct observation has identified 110 species so far (2008–2012) and covers a range of up to 12 km around the platform, depending greatly on the species' size and visibility in relation to weather. Information on behavioural responses can therefore be gathered for many species by using evenly balanced sample counts in different viewing directions alternately towards and away from the windfarm. The results show great variety, ranging from avoidance of a windfarm (e. g. little gull, ◘ Fig. 12.21) to complete disregard (large gulls of the genus *Larus*), while there is so far no evidence of attraction during daylight. A rough halving of diversity among migrants counted in sectors directed towards *alpha ventus*, however, may indicate a dominance of avoidance behaviour among all species involved.

Effort is currently underway to identify species-specific reaction thresholds to *alpha ventus*. Some species, though recorded in lower numbers in sectors directed towards the windfarm, seem to show considerable day-to-day variation in reactions. Whether this is driven by individual decisions (based on factors such as feeding conditions in or around the windfarm) and/or is passively induced by external factors (e. g. strong winds limiting manoeuvrability) is something that needs further investigation at species level. The findings provide a basis for modelling the cumulative effects on birds crossing not just a single windfarm but the extensively industrialised German Bight as a whole.

12.6 Perspectives

Ornithological research at *alpha ventus* has so far answered basic questions about what may happen to migratory birds crossing the open sea when wind turbines increase in number. However, results from *alpha ventus* cannot be transferred to other windfarms without further evaluation. Future research needs to address the factors influencing the probability of detecting migrating birds at sea and potential differences between localities. In particular, the following questions remain to be answered:

- How does migration intensity at sea vary between different geographical regions (e. g. North Sea versus the Baltic region)?
- How often and under what circumstances do night-migrating birds accumulate in windfarms and to what extent is this affected by turbine activity?
- How can avian collisions at offshore wind turbines be measured directly and with sufficient coverage over extended time periods?

- How does the risk of collision change with turbine arrangement, size and activity?
- How do the combined (cumulative) effects of artificial lights mounted on construction vessels, offshore platforms and wind turbines affect collision rates in nocturnal migrants?
- Will avian collisions at offshore wind turbines or the distortion of flight routes by windfarms reach an order of magnitude that affects populations?
- What mitigation measures are possible with respect to the design of wind turbines and their illumination? Is temporary shutdown of wind turbines an effective and feasible mitigation strategy?

Answering these questions is pivotal in order to provide policymakers, governmental agencies, and industry with the information needed to improve planning, approval and mitigation processes. Unbiased, reliable information is also a powerful tool for promoting public acceptance of renewable energy technologies in the 'green versus green' debate.

Automated stationary surveillance of birds in the vicinity of wind turbines and in conjunction with information derived from large-scale radar networks has the potential to create baseline data that can be effectively used to answer most of the above questions. However, without accurate quantification of bird migration at species level, in-depth understanding of migratory connectivity, and a method to quantify actual collision events, we will not be able to predict how human activities will impact on bird populations. It is likely that bird mortality caused by collisions with offshore wind turbines will rarely exceed mortality caused by otherwise accepted human-caused impacts (such as agriculture, feral cats, and bird strike at windows and power lines). Anthropogenic mass mortality may nonetheless be more frequent and pronounced in marine environments than on land (Newton 2008). There is consequently no argument at present why offshore windfarms should not be carefully planned or restricted in size to minimise additional pressure on birdlife.

Regulations regarding mass bird migrations (Incidental Provision 21)

Anika Beiersdorf

Under German law, offshore windfarms can only be approved if harmful effects on the marine environment are sufficiently unlikely. The Marine Facilities Ordinance (Seeanlagenverordnung/SeeAnlV) is the legislation governing the licensing of offshore windfarms in the German Exclusive Economic Zone (EEZ). One of the reasons for refusing approval is if a windfarm poses a threat to the marine environment, and notably to migrating birds (Section. 5 (6) (2), SeeAnlV) (◘ Fig. 12.22).

Based on this legislation, offshore windfarm approvals granted by BSH prescribe standard procedures for the event of mass migration across an offshore site. In such an event, windfarm operators must comply with Incidental Provision 21 in the windfarm approval document (see Chap. 1).

This states that if a mass migration event is likely, operators must ensure that data is collected, in particular on any incidence of bird strike during the mass migration. The findings must be made available to BSH within one week after the event. BSH can require turbines to be fitted with deterrent devices or to be temporarily shut down in the event of a mass migration.

This makes it essential to assess avian collision risk at offshore installations. The number of migratory birds in the rotor-swept zone has to be determined and set in proportion to overall migration rates measured in and around the windfarm. Ornithological research at *alpha ventus* has shown what pressures may act on migratory birds crossing the open sea. Further research is now needed to develop forecasting models for a real-time prediction of such events and to consider efficient avoidance and mitigation measures.

◘ **Fig. 12.22** A flock of migrating eider ducks (*Somateria mollissima*) (photo: Mathias Putze).

Literature

Aumüller R, Boos K, Freienstein S, Hill K, Hill R (2011). Beschreibung eines Vogelschlagereignisses und seiner Ursachen an einer Forschungsplattform in der Deutschen Bucht. Vogelwarte 49:9–16

Aumüller R, Boos K, Freienstein S, Hill K, Hill R (2013). Weichen Zugvögel Windenergieanlagen auf See aus? Eine Methode zur Untersuchung und Analyse von Reaktionen tagsüber ziehender Vogelarten auf Offshore-Windparks. Vogelwarte 51:3–13

Berthold P (2001). Bird migration. A general survey, 2nd edn. Oxford University Press, Oxford, New York

Bruderer B (1997). The study of bird migration by radar. Part 1: The technical parts. Naturwissenschaften 84:1–8

Christensen TK, Hounisen JP, Clausager I, Petersen IK (2004). Visual and radar observations of birds in relation to collision risk at the Horns Rev offshore windfarm. Annual status report 2003. National Environmental Research Institute, Denmark

Clarke WE (1912). Studies in bird migration. Vol. 2, Gurney & Jackson, London

12

Cooper BA, Day RH, Ritchie RJ & Cranor CL (1991). An improved marine radar system for studies of bird migration. J. Field Ornithol. 62: 367–377

Desholm M & Kahlert J (2005). Avian collision risk at an offshore windfarm. Biol. Lett. 1:296–298

Dierschke J, Dierschke V, Hüppop K, Hüppop O, Jachmann KF (2011). Die Vogelwelt der Insel Helgoland. OAG Helgoland, Helgoland

Drewitt AL, Langston RHW (2006). Assessiing the impacts of wind farms on birds. Ibis 148: 29-42

Drost R (1960). Über den nächtlichen Vogelzug. Proc. XII. Int. Ornithol. Congr.:178–192

Eastwood E (1967). Radar Ornithology. London: Methuen

Farnsworth A (2005). Flight calls and their value for future ornithological studies and onservation research. Auk 122: 733–746

Gätke H (1895). Heligoland as an ornithological observatory. The results of fifty year experience. Edinburgh University Press, David Douglas, Edinburgh

Gauthreaux, SA & Belser, CG (2006). Effects of Artificial Night Lighting on Migrating Birds. In: Rich C & Longcore T (eds.): Ecological Consequences of Artificial Night Lighting. Island Press, London. pp. 67–93.

Hansen L (1954). Birds killed at lights in Denmark 1886–1939. Vidensk. Medd. Dan. Naturh. Foren. 116:269–368

Hill R & Hüppop O (2007). Methods for investigating bird migration. In: Morkel L, Toland A, Wende W, Köppel J (Edts.): Conference Proceedings 2nd Scientific Conference on the Use of Offshore Wind Energy by the Federal Ministry: 143–152.

Hill R & Hüppop O (2008). Birds and Bats: Automatic Recording of Flight Calls and their Value for the Study of Migration. In: Frommolt KH, Bardeli R & Clausen M: Computational bioacoustics for assessing biodiversity. BfN-Skripten 234: 135–141

Hüppop O, Dierschke J, Wendeln H (2004). Zugvögel und Offshore- Windkraftanlagen: Konflikte und Lösungen. Ber. Vogelschutz 41: 127–218

Hüppop O & Hilgerloh G (2012). Flight call rates of migrating thrushes: effects of wind conditions, humidity and time of day at an illuminated offshore platform. J. Avian Biol. 43: 85–90.

Hüppop O & Hill R (2007). Bird migration over the North Sea. In: Morkel L, Toland A, Wende W, Köppel J (Edts.): 2nd Scientific Conference on the Use of Offshore Wind Energy by the Federal Ministry for the Environment 20. and 21. February 2007 in Berlin Conference Proceedings: 35–40.

Langston RHW & Pullan JD (2003). Windfarms and birds: an analysis of the effects of windfarms on birds, and guidance on environmental assessment criteria and site selection issues. Report T-PVS/Inf (2003) 12, by BirdLife International to the Council of Europe, Bern Convention on the Conservation of European Wildlife and Natural Habitats. RSPB/ BirdLife in the UK

Martin GR (2011). Understanding bird collisions with man-made objects: a sensory ecology approach. Ibis 153:239–254

Neumann R, Kube J, Liechti F, Steuri T, Wendeln H, Sordyl H (2009). Entwicklung einer Methode zur automatischen Quantifizierung des Vogelzuges im Bereich von Offshore-Windparks und der Barrierewirkung der technischen Anlagen für den Vogelzug mittels fixed beam Radar. Abschlußbericht zum Forschungsvorhaben des Bundesministeriums für Umwelt, Naturschutz und Reaktorsicherheit (FKZ 0327632)

Newton I (2008). The migration ecology of birds. Academic Press, London

Schulz A, Kube J, Kellner T, Schleicher K, Sordyl H (2009). Entwicklung und Einführung eines automatischen Erfassungssystems für die Ermittlung des Vogelschlages unter Praxisbedingungen auf FINO2. Abschlußbericht. Forschungsvorhaben des Bundesministeriums für Umwelt, Naturschutz und Reaktorsicherheit (FKZ 0327560). Neu Broderstorf

Schulz A, Kulemeyer C, Röhrbein V, Coppack T (2011). The extent of phototactic attraction of night-migrating birds passing an illuminated steel mast in the western Baltic Sea. International Conference on wind energy and wildlife impacts, Trondheim, Norway. NINA Report 693:102

Taylor PD, Brzustowski JM, Matkovich C, Peckford ML, Wilson D (2010). radR: an open-source platform for acquiring and analyzing data on biological targets observed by surveillance radar. BMC Ecology 10: 22.

Vaughan R (2009). Wings and rings. A history of bird migration studies in Europe. Isabelline Books, Penryn, Cornwall

Winkelman JE (1992). De invloed van de Sepproefwindcentralete Oosterbierum (Fr.) opvogels: 1: aanvaringsslachtoffers. DLO-Instituutvoor Bos- en Natuuronderzoek, Arnehem. Rinrapport 92/2

Zaugg S, Saporta G, van Loon E, Schmaljohann H, Liechti F (2008). Automatic identification of bird targets with radar via patterns produced by wing flapping. Journal of the Royal Society Interface 26:1041–1053

Marine mammals and windfarms: Effects of *alpha ventus* on harbour porpoises

Michael Dähne, Verena Peschko, Anita Gilles, Klaus Lucke, Sven Adler, Katrin Ronnenberg, Ursula Siebert

Federal Maritime and Hydrographic Agency,
Federal Ministry for the Environment, Nature Conservation and Nuclear Safety (Eds.)
Ecological Research at the Offshore Windfarm alpha ventus,
DOI 10.1007/978-3-658-02462-8_13, © Springer Fachmedien Wiesbaden 2014

13.1 Introduction

The harbour porpoise (*Phocoena phocoena*) is the only native species of toothed whales in the German North Sea. Two native species of phocid seals occur: the harbour seal (*Phoca vitulina*) and the grey seal (*Halichoerus grypus*). These species are of special importance for estimating the environmental effects that the construction, operation and decommissioning of offshore windfarms may have on marine wildlife. Other cetacean species like minke whale (*Balaenoptera acutorostrata*) and white-beaked dolphin (*Lagenorhynchus albirostris*) are encountered on a regular basis in the German North Sea, however, the effects that German windfarms may have cannot be sufficiently evaluated due to low occurrence rates. While this chapter concentrates primarily on harbour porpoises, it provides a brief summary concerning the species that need to be considered in the North Sea.

The possible effects of offshore windfarms on marine mammals can be numerous and are described in Sects. 13.1.2 and 13.1.3. These effects are required to be monitored in a properly designed environmental impact monitoring programme such as prescribed by BSH Standard for Environmental Impact Assessment (StUK). The study presented here was conducted in addition to the StUK3 study scope and focused on estimating the impact of pile driving on harbour porpoises in a larger area during the construction period of *alpha ventus*, and monitoring this area during the operation period (BSH 2007).

13.1.1 Key Species

Harbour porpoise (*Phocoena phocoena*)

Harbour porpoises (◧ Fig. 13.1) occur circumpolar in temperate to boreal waters in the northern hemisphere. They can reach a weight of 45 to 75 kg (maximum ~ 90 kg for pregnant females) and a length of 150 to 165 cm (maximum ~ 185 cm) and have a maximum life span of around 25 years. Porpoises feed on small to medium sized mostly pelagic, but also demersal and benthic, fish species.

Within German waters three populations are currently distinguished: 1) the eastern North Sea, including the Skagerrak and the northern part of the Kat-

tegat, 2) the Kattegat, the Belt Sea and Western Baltic, and 3) the Baltic Proper. The most recent population assessment conducted on the European Atlantic continental shelf found no difference in abundance between 1994 and 2005 surveys. However, a marked difference in harbour porpoise distribution has been shown from estimated higher densities in northern areas in 1994 towards a more southerly hotspot in 2005 (Hammond et al. 2002, 2013). SCANS II estimated an abundance of 375,358 individuals (95 % confidence interval (CI): 256,304–549,713) in the European North Atlantic in 2005 (Hammond et al. 2013). Between 15,394 (CV = 0.33) and 55,048 (CV = 0.30) porpoises occur in the German North Sea depending on season (Gilles et al. 2009).

Porpoises are under pressure from incidental bycatch, prey depletion, pollution, anthropogenic noise and probably climate change in cold waters. For noise exposure during pile driving, a criterion of a sound exposure level (SEL) of 160 dB re µPa²s outside of a circle of 750 m radius has been implemented into the licensing process in Germany, based on the measured level for a temporary threshold shift in harbour porpoises of 164.3 dB re µPa²s for impulsive low frequency sounds (Lucke et al. 2009). To protect marine mammals from loud pile driving noise, mitigation strategies during pile driving were strongly promoted (▶ see Chap. 16).

Surveys conducted in the German North Sea between 2002 and 2006 showed that porpoise occurrence varied between seasons (Gilles et al. 2009). The highest abundance was estimated for the early summer months with highest densities occurring at the Sylt Outer Reef, north-east of the EEZ. Harbour porpoise density in the southern German Bight increased from 2004 onwards, at first mainly in spring (area D in Gilles et al. 2009), and from 2008 onwards also in summer (Gilles et al. 2011, Dähne et al. 2013a). A general increase in porpoise abundance in the southern North Sea is also reported by neighbouring countries (e.g. Scheidat et al. 2012, Hammond et al. 2013). At the Borkum Reef Ground (BRG), south-west of the EEZ near the *alpha ventus* windfarm, a hot-spot, i. e. a discrete area of particularly high harbour porpoise density, was observed over several study years. This area probably serves as one of the two key foraging areas in German waters (Gilles et al. 2009). Seasonal variation was also

■ **Fig. 13.1** Harbour porpoise (*Phocoena phocoena*) (photo: Katharina Brandt / Dolfinarium Harderwijk).

■ **Fig. 13.2** (a) harbour seal (*Phoca vitulina*), (b) grey seal (*Halichoerus grypus*) (photo: (a - b) Michael Dähne / ITAW).

observed in the Netherlands, with the highest densities in winter and spring, and the lowest in summer (Scheidat et al. 2012).

Harbour seal (*Phoca vitulina*)

Harbour seals (■ Fig. 13.2a) inhabit the ice-free coastal waters of the northern hemisphere with four marine and one freshwater sub-species. Males are larger than females in terms of body length (180 to 150 cm) and weight (maximum of 130 to 105 kg). Fur coloration is often irregular in grey and brown patterns, and head shape is typically round with a dog-like snout. In Germany, harbour seals form colonies on the sand banks and beaches of the Wadden Sea. They are less common in the German Baltic Sea, but abundant around the Danish islands. Worldwide population is estimated around 350,000 to 500,000 individuals. Since cessation of seal hunting in Germany, numbers have increased steadily from ~3,000 to 16,424 aerially counted individuals in Schleswig-Holstein, Lower Saxony and Hamburg (TSEG 2013) with interruptions by two large epidemics of phocine distemper virus causing die-offs in 1988 and 2002.

Harbour seals are opportunistic feeders on demersal fish, using their vibrissae to detect vortices caused by moving fish in the water. They use foraging trips along the coastal shelf to hunt for gadoids and flatfish, and can be sighted at large distances of ~ 150 km offshore. In the German North Sea, seal pups are born on the haul-outs from May to July and are nursed for around four weeks. They can reach a maximum age of 30 to 35 years.

Grey seal (*Halichoerus grypus*)

Grey seals (◘ Fig. 13.2b) are one of the most commonly occurring seal species in the northern hemisphere. They are larger in size and weight than harbour seals and are easily distinguishable by their cone shaped head with no visible forehead in comparison to the dog shaped head of *Phoca vitulina*. Grey seals of the western and eastern Atlantic have different body sizes. Eastern Atlantic males have a mean body length of 200 cm and weigh up to 310 kg. They show a strong sexual dimorphism with the females being much smaller at 180 cm length and 155 kg in weight, and showing differences in coloration and head shape. Fur coloration in males is often dark brown to dark grey or black, while females are light grey to off-white with dark spots. New born pups show a white-coloured dense lanuginose fur for approximately one month after birth.

Grey seals and harbour seals occur on the same haul-out sites, but for grey seals the islands Norderney, Helgoland and Amrum are currently the major haul-out sites in the North Sea. In the Baltic Sea, large colonies are found on the Danish and Swedish coastlines. In the German Baltic Sea, a haul-out has been established recently within the Greifswalder Bodden. Increasing numbers of animals are observed in both the German North and Baltic Seas. Grey seals have a life span of 35 to 40 years.

13.1.2 Possible effects of windfarm construction

Offshore windfarms have the potential to affect marine mammal populations. A number of different factors must be considered. For harbour porpoises, the threat considered most important at the moment is the influence of impulsive (pile driving) or continuous (vibratory piling) noise during the construction phase (Madsen et al. 2006). Pile driving noise is potentially loud enough to cause a temporary or permanent hearing loss in porpoises and seals very close to the piling site. Temporary hearing loss (temporary threshold shift / TTS) has been recorded in a porpoise in human care at a sound exposure level (SEL) of 164 dB re µPa²s and a peak to peak pressure level ($L_{peak-peak}$) of 199.3 dB re µPa for impulsive

airgun noise (Lucke et al. 2009). Hence, the maritime and environmental agencies in Germany propose a precautionary level of an SEL of 160 dB re µPa²s and L_{peak} of 190 dB re µPa at 750 m distance to the piling site. It must be noted that this is a single impulse criterion and, from the authors perspective, does not cover multiple exposures likely to be experienced during windfarm construction with 11,383 to 25,208 hammer strokes in ~ 2 s intervals per foundation (not counting piling breaks) consisting of 3–4 smaller piles as an example from *alpha ventus* in all circumstances. However, all marine mammals in the close vicinity of the construction site are potentially subject to temporary hearing loss, increased stress levels, an avoidance reaction and therefore temporary habitat loss and a flawed equilibrium sense during pile driving. Additional noise is created due to increased ship traffic for maintenance and possible excavation work prior to installing the foundations. However, this impact continues during the operation period and hence should be considered as an operational effect in general.

13.1.3 Possible effects of windfarm operation

Effects of the operational period that need to be considered can be either noise effects or effects due to alteration of the habitat where foundations were erected. Noise effects may include displacement or an attraction of the animals due to the noise emitted. For seal species, behavioural reactions can range around hundreds of meters (Tougaard et al. 2009b). Due to the nature of operation noise as a continuous noise source at low frequencies (< 500 Hz), masking effects on communication sounds should also be considered. This is especially true for the two seal species in German waters as their hearing abilities are much better in low frequencies than those of harbour porpoises and they vocalise in adjacent frequency bands. Porpoises and seals may habituate to these noise impacts, i. e. they may change their behaviour over time to adapt to the new situation. Habituation cannot be seen a priori as a 'bad' or 'good' effect, but must be put into context. Habituation to pingers for instance could be a negative effect, because it may prevent the pinger from having its intended effect of mitigating by-catch. However, a certain amount of habituation to pingers may be a positive effect as it might enable porpoises to use their natural habitat to a better degree, as long as it does not result in more porpoises being by-caught or being otherwise negatively affected. Testing for the described noise effects is very difficult as suitable monitoring methods for the seals most likely affected are not available at present. What is certain is that the operation of offshore windfarms over a longer period of time most likely has a small spatial scale impact compared with the effect of the short-term construction period, which has an impact over vast distances. However, investigations around the Nysted windfarm (Denmark) for a longer period showed a negative long-term effect of the construction period (Teilmann & Carstensen 2012).

Additional potentially positive effects on the marine environment during operation include the creation of artificial reefs that have an impact on prey species abundance and species diversity at the respective windfarm site, unless, that is, operational noise deters fish. Studies have speculated that windfarms may be an attraction for porpoises due to the increased prey abundance, however the effect is most likely small (Diederichs et al. 2008) although experiments on fish diet suggest that fish closer to the Thorntonbank windfarm had a fuller stomach than further away (Derweduen et al. 2012).

An artificial reef effect may also be the reason for increased porpoise detection rates observed at Egmond aan Zee, the first offshore windfarm in the Netherlands (Scheidat et al. 2011). However, the windfarm area is also closed for ship traffic and fisheries, and may thus have the effect of a generally 'noise-reduced' area due to reduced shipping activity. It may also serve as a protected zone for pelagic and benthic fish species, thus further increasing food availability.

13.2 Methods

13.2.1 Assessing presence, absence and abundance

Animal abundance is an essential factor in assessing anthropogenic impacts on biota and to inform con-

servation and management. Further, understanding of population dynamics and spatial distribution, on both an annual and seasonal level, is important to establish a baseline for evaluation of the effects of human disturbance. Effect monitoring at offshore windfarms should preferably include a survey immediately before construction activities begin to ensure that the baseline is still relevant. This may lead to a Before-After-Control-Impact (BACI) survey design. However, in practice it is difficult to select comparable control sites. In the context of windfarms, where the likely source of impact is noise – an impact that propagates and diminishes gradually over large distances – a gradient sampling design (e.g. by static acoustic monitoring) could be more useful. Preferably, more than one method should be selected as effects of a stressor may manifest in different categories: Single/short term, single/long-term, multiple/short term and multiple/long-term.

In general, marine mammals are difficult to study as they spend most of their lives in or under water. A standard method for estimating cetacean abundance is line transect distance sampling, where a ship or an aircraft surveys along a series of transects. The sample density is then extrapolated to the entire survey area. Seals are commonly counted by aerial photography when they come ashore to breed or to moult.

For the *alpha ventus* offshore windfarm, aerial surveys and static acoustic monitoring (SAM) were carried out in parallel. In addition, shipboard line transect sampling following SCANS II standards (Hammond et al. 2013) was used to test whether effects could be detected at a small scale in the vicinity of the windfarm. Furthermore, a towed hydrophone system capable of recording porpoise echolocation clicks during night and day, in bad weather conditions and independent of an observer error was deployed during the ship surveys. In the following, only results from the aerial surveys and SAM are presented.

Visual surveys and SAM are established methods for porpoise monitoring. While aerial surveys can also be used to estimate seal abundance (Herr et al. 2009), they fail when the effects of a windfarm on seals need to be assessed. Mark-recapture techniques may be used to estimate abundance where individuals can be recognised through natural

a   b

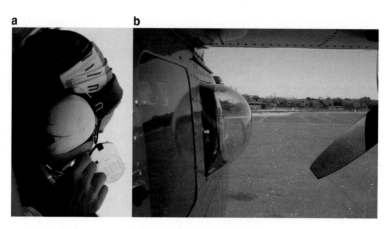

■ **Fig. 13.3** (a) A marine mammal observer uses an inclinometer to measure the declination angle towards a porpoise sighting for calculation of the perpendicular distance to the sighting. (b) Bubble window of a high-winged aircraft for observation beneath the plane (photo: ITAW).

markings or tags (e. g. photo-identification studies of bottlenose dolphins in the Moray Firth, UK). Platforms of opportunity, monitoring of strandings, acoustic data loggers, towed hydrophone surveys, fixed-point surveys and telemetry are also widely used methods to infer marine mammal presence, habitat use, behaviour, population health and life history, but cannot be used to estimate abundance.

13.2.2 Aerial line transect surveys

Line transect sampling from ship or aircraft is commonly used to estimate cetacean density (e.g. Gilles et al. 2009, Scheidat et al. 2012, Hammond et al. 2013). The density of the target species is estimated by surveying along a series of pre-designed transects; this sample density is then extrapolated to the entire survey area. The method thus provides an estimate of the number of animals in a defined area at a particular time or over a certain period. Between August 2008 and October 2012, 19 aerial line transect surveys were conducted to assess the density and distribution of harbour porpoises. The survey area (10,934 km^2) comprised a 60 km radius around *alpha ventus* to study the large-scale effect of windfarm construction and operation on porpoise presence as well as spatio-temporal trends in distribution (■ Fig. 13.3). In order to provide equal coverage probability, i. e. where every point in the survey area has the same theoretical probability of being sampled, a parallel transect layout was selected using DISTANCE 5.0 (Thomas et al. 2010), with 15 transects spaced seven km apart (total tran-

sect length 1,780 km). All transects were surveyed within two days.

Aerial surveys were flown at 90 to 100 kn (167 to 185 km/h) at an altitude of 600 ft (183 m) in a Partenavia P68, a twin-engine, high-wing aircraft equipped with two bubble windows to allow scanning directly below the plane. The survey team, consisting of two observers and one data recorder (navigator), conducted surveys only in sea conditions Beaufort 0 to < 3 and with visibility > 5 km. Environmental conditions were recorded at the beginning of each transect and updated to document every change. Sighting conditions were classified as good, moderate or poor based primarily on sea state, water turbidity and glare. The observers recorded the perpendicular distance to each sighting (■ Fig. 13.3) as well as data on group size, group composition (e.g. mother-calf pair) and behaviour. Using the distance data, a detection function was fitted strip which was then used to estimate the effective strip width; this corrects for animals further away from the transect that were missed by observers. In order to account for porpoises missed because they were under water (availability bias) or because they were on the surface but not seen by the observer (perception or observer bias), the racetrack data collection method was used. In this method, the aircraft circles back over part of a previously surveyed transect and duplicates are determined probabilistically from a model incorporating data such as speed of movement of the target species. Detailed field and analysis protocols are described in Gilles et al. (2009).

13.2.3 Static acoustic monitoring (SAM)

Harbour porpoises use their echolocation as an active sense almost constantly for spatial orientation and foraging with only a few brief breaks. These high-frequency sounds are probably also used for communication. Porpoise echolocation signals, called clicks, are narrow band sounds with peak frequencies between 120 and 140 kHz which have strong tonal characteristics in a frequency band where few other sounds are produced naturally. This makes it feasible to automatically detect porpoise echolocation clicks using high frequency hydrophones either in surveys employing towed hydrophones or in static acoustic monitoring (SAM) deploying automated underwater click detectors. Using SAM it is possible to find variations in presence/absence patterns, diel cycles and behaviour. The most commonly used device in Europe is the C-POD (◘ Fig. 13.4). C-PODs detect all tonal sounds within a frequency range of 20 to 160 kHz, have a linear relationship between received peak to peak sound pressure level ($L_{peak-peak}$) and registered click amplitude from 80 to 130 kHz and have a detection threshold of 114.5 dB reµ Pa± 1.2 dB at 130 kHz (Dähne et al. 2013b). Although comparability in sensitivity between devices is much better than for T-PODs (Verfuß et al. 2013), it is advisable to perform test tank calibration at regular intervals to guarantee that devices do not change within the course of a study. The study devices can run on 10 D-Cells for up to six months continuously in low noise conditions and log recorded click timestamp, peak to peak amplitudes as uncalibrated 8-bit digits, parameters of the envelope, frequency and bandwidth. The C-POD not only logs porpoise echolocation clicks, but also shipping noise (cavitation and echo sounders), background noise during periods with high seas, high frequency components of broadband noise like pile driving, and vocalisations of other biota like crustaceans. Hence, it is necessary to filter the recorded clicks to find sequences from echolocating harbour porpoises.

Regular patterns in the interval between two successive clicks, as well as in the amplitudes are used by an automated routine to find 'click trains' –

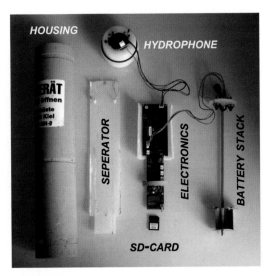

◘ **Fig. 13.4** A C-POD and its major components (photo: ITAW).

segments of echolocation click sequences. All detected click trains are ranked by the probability of being indeed caused by echolocating harbour porpoises. Classified click sequences can then be used to calculate different parameters: 'Detection positive' time periods (minutes, ten minutes, hours and days) – periods containing at least one sequence classified as a narrow band high frequency (i. e. 'porpoise-like') click train – and waiting times (WT), being the interval between two successive porpoise detections separated by at least ten minutes without a detection. Detection positive 10 minutes (dp10 min) and detection positive hours (dph) can be averaged over a longer time span and give a measure of presence of porpoises during a specific period (construction/operation/decommissioning). These metrics can be used to show that porpoises use an area less frequently for instance during pile driving (Brandt et al. 2011). Waiting times, by way of contrast, provide a measure for absence of porpoises and can be used to assess how long porpoises are displaced during and after pile driving (e.g. Tougaard et al. 2009a).

SAM has been used in a number of studies to assess the impact of the construction and operation of offshore windfarms on cetaceans. Continuous monitoring of a specific site allows conclusions to be drawn either about a relatively short period

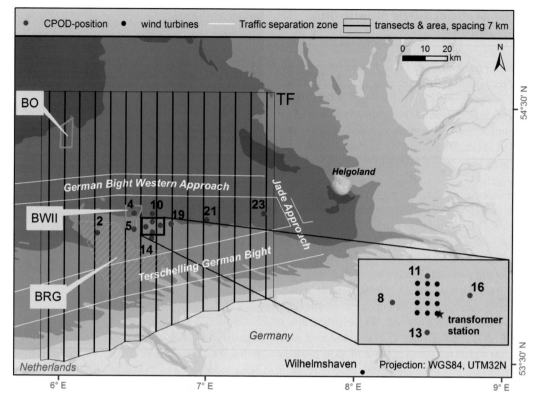

☐ **Fig. 13.5** Map of the study area around *alpha ventus*. BO: BARD Offshore 1 offshore windfarm, BRG: Borkum Reef Ground Natura 2000 SCI, BWII: Trianel windfarm Borkum offshore windfarm (first stage), TF: line transect layout for aerial surveys for the *alpha ventus* test site.

like the construction or about long time periods like the operation of an offshore windfarm in comparison to baseline data. It is not yet possible to calculate abundance from SAM data, although methods have been suggested (Kyhn et al. 2012). For other, more traditional methods, like aerial or ship-based surveys, it is much more difficult to establish a direct comparison of porpoise abundance data from impact studies concerning effects that are either on a small spatial scale (operation period) or short termed (construction period). But they allow acoustic registration rates of click detectors to be supplemented with abundance/density of animals to estimate impact on population scales as suggested by Gilles et al. (2009).

13.3 Results

13.3.1 Construction effects on harbour porpoises

Before, during and after the construction of the *alpha ventus* test site, comprehensive monitoring was carried out using aerial and ship-based standard line transect distance sampling surveys along with continuous SAM on twelve measuring positions within the StUKplus research project (☐ Fig. 13.5). Originally 21 positions were planned, but the study design had to be reduced due to numerous losses in areas of intense fishery efforts. This study was conducted at a larger scale than the compulsory environmental monitoring carried out in accordance with the StUK3 standard. Both monitoring approaches were designed to supplement each other, with the goal of providing the best available data basis for investigating possible short and

Table 13.1 Significance of GAM models of dp10 min/h from 2008 to 2010 (from Dähne et al. 2013a, © IOP Publishing Ltd. CC BY-NC-SA). The intercept describes the mean of the model and can be compared to intercepts of predictor variable like 'pile driving' to estimate whether an impact raises or lowers the modelled mean. n – number of samples, p – significance level, expl. dev. – explained deviance.

Position	Distance to piling site *min – max* (km)	n	Intercept	Intercept pile driving	effect	p pile driving	p year	p month	p hour	expl. dev.
2	25.2 – 26	6848	0.99	n.s.	n.s.	n.s.	n.s.	< 0.001	< 0.001	8.23 %
4	8 – 10.8	13315	0.88	−0.42	−	< 0.001	< 0.001	< 0.001	0.025	10.87 %
5	7.4 – 9.8	12039	−0.66	−1.24	−	< 0.001	< 0.001	< 0.001	< 0.001	17.08 %
8	2.3 – 4.6	12838	0.42	−1.36	−	< 0.001	< 0.001	< 0.001	< 0.001	10.54 %
10	3.0 – 4.2	5602	1.08	−0.61	n.s.	n.s.	< 0.001	< 0.001	< 0.001	19.84 %
11	0.5 – 2.5	14226	0.00	−1.16	−	< 0.001	< 0.001	< 0.001	< 0.001	13.92 %
13	2.3 – 4.7	12823	−0.55	−0.86	−	< 0.001	< 0.001	< 0.001	< 0.001	6.46 %
14	4.5 – 7.0	12846	2.22	−0.81	−	< 0.001	< 0.001	< 0.001	< 0.001	8.90 %
16	2.5 – 4.5	11286	0.76	−1.67	−	< 0.001	< 0.001	< 0.001	0.003	20.07 %
19	7.2 – 9.2	14970	1.28	−1.51	−	< 0.001	< 0.001	< 0.001	0.095	16.81 %
21	23 – 25	7283	−1.81	0.25	+	0.005	< 0.001	< 0.001	< 0.001	13.81 %
23	48.7 – 50.5	9406	−0.62	−0.54	+	< 0.001	n.s.	< 0.001	< 0.001	3.84 %

long-term impacts of the *alpha ventus* windfarm on porpoises.

Spatial displacement

For each C-POD position independently, generalised additive models (GAMs) – a form of non-linear regression analysis – were constructed to assess the impact of pile driving on porpoise presence. Each model was fitted using the variable dp10 min/h as the response or dependent variable. Pile driving (yes/no), year, month and hour of day were used as predictor or independent variables. The outcomes of these models are summarised in ❏ Table 13.1. Most of the positions show diurnal, seasonal (monthly) and annual differences. A low in porpoise registrations was recorded each year from the end of April to the beginning of August.

Ten of the twelve models showed a significant impact of pile driving – eight positions close to *alpha ventus* (within 11 km of the piling site) showed negative impacts, while two at distances of 23 to 50 km respectively showed a positive correlation (raised dp10 min/h during pile driving). There were

no positions between 11 and 23 km. This effect of pile driving in the closer vicinity of the windfarm is a combined effect of pile driving itself and displacement caused by deterrence devices. The spatial distribution pattern recorded in two aerial surveys three weeks before and exactly during pile driving at *alpha ventus* points to a strong avoidance response in harbour porpoises within 20 km distance of the noise source (Dähne et al. 2013a). This confirms findings of previous studies in the Danish North Sea which showed a significant reduction in detection rates up to 17.8 km (Brandt et al. 2011) or even above 21 km (Tougaard et al. 2009a).

To investigate at which distances a displacement effect may occur, two other models of the dp10 min were taken into account (❏ Fig. 13.6). In these models, the distance to the actual pile driving site was included as an independent variable, and month and position were included as random variables. Their effect in terms of seasonal and geographic variation within the model was hence accounted for to provide an unbiased estimate for the distance effects.

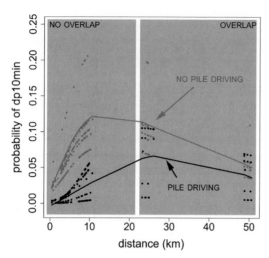

● **Fig. 13.6** Extent of the displacement effect (adapted from Dähne et al. 2013a, © IOP Publishing Ltd. CC BY-NC-SA). The chart shows the predictions of two models using the dp10 min determined for 2008–2011 (one with pile driving in black and the other without in grey) for the months when pile driving was conducted in the years 2008 and 2009. A smoothing function with five degrees of freedom was used to visualise the general pattern in both predictions.

The predictions provided by these models show no overlap up to 11 km from the pile driving site and overlap from 23 km onwards. A geographical gradient shows that detection rates are generally much lower in close vicinity of the windfarm regardless of pile driving activity. However, the overlap from 23 km shows that displacement of porpoises due to pile driving gradually fades out – some animals may still react, while others do not.

Temporal displacement

To assess how long porpoises were displaced during pile driving, we analysed waiting times (WT) (● Fig. 13.7). The WT directly after the pile driving stopped ranged from 81 minutes (position 2) to 1,444 minutes (position 11) in median with an overall median of 1,008 minutes (16.8 h) and a maximum of 8,468 minutes (141.1 h, position 13). For comparison: median WT not affected by pile driving in 2008 to 2011 ranged between 46 (0.8 h) to 67 (1.1 h) minutes for all stations.

The first WT after individual pile driving for each pile gradually decreased towards the end of the pile driving activities at *alpha ventus* in 2009.

Models showed a significant correlation between pile driving duration and the first WT independent of the equally significant seasonal variation. Hence, the longer pile driving was carried out, the longer porpoises were displaced. The most likely explanation is that porpoises are displaced during pile driving and at a certain point, stop to further move away when their behavioural reaction threshold is no longer triggered. During shorter piling events, the available time might not be sufficient for porpoises to have reached this distance. More details of the above analyses are provided in Dähne et al. (2013a).

Use of acoustic harassment devices

A potentially detrimental radius around pile driving sites can be defined using the limit for a temporary impairment of the hearing system of porpoises (Lucke et al. 2009). The defined threshold value for impulsive underwater noise (see Sect. 13.3.1) is intended to prevent harbour porpoises outside 750 m from experiencing a TTS from single exposures. However, this level was exceeded in most cases at *alpha ventus* (Betke & Matuschek 2011) and porpoises must be excluded from inside the 750 m radius. To achieve the latter, pingers – acoustic devices to scare porpoises away from fishing nets – were used to scare porpoise from the direct vicinity of the piling site. Additional use was made of seal scarers, developed as a potential mitigation measure at aquaculture sites to prevent seal predation. Seal scarers proved to be very effective to deter porpoises. The device emitted louder tonal sounds to extend the range of displacement of the pingers before pile driving was conducted. The range of porpoise displacement through seal scarers has been etimated at up to 7.5 km in the North Sea (Brandt et al. 2012) and 2.4 km in the Baltic Sea (Brandt et al. 2013). These far-reaching effects highlight the additional effect of seal scarers used as a measure to reduce porpoise occurrence within a potentially detrimental radius before pile driving starts. Hence, seal scarers cannot replace the use of noise mitigation measures such as bubble curtains as a method of source mitigation.

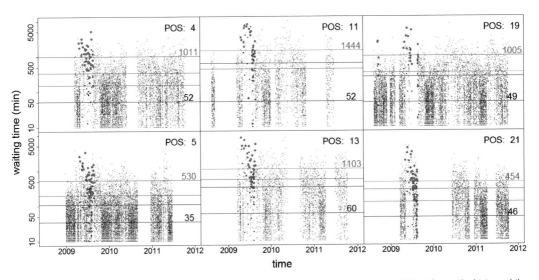

■ **Fig. 13.7** Waiting times at six selected C-POD positions. Grey dots indicate the waiting times (WT) without pile driving, while red dots indicate the first WT, ending with the first registration after pile driving stopped. Blue dots indicate the successive (second) WT. Coloured lines indicate the median of the corresponding dots and the light grey line indicates the median + standard deviation (sd) of WT without pile driving. The y-axis is log-scaled. The first WT gradually decreased towards the end of the pile driving activities at *alpha ventus* in 2009 (adapted from Dähne et al. 2013a, © IOP Publishing Ltd. CC BY-NC-SA).

13.3.2 Operation effects on harbour porpoises

Distribution and abundance

Between 2008 and 2012, 19 dedicated aerial surveys for harbour porpoises were conducted, covering a total effort of 23,338 km and resulting in 1,999 sightings of harbour porpoise groups. The number of sighted individuals totalled 2,392, including 107 calves. While calves were sighted in the months April to September, most were sighted in the summer months, June to August. No distinct breeding area was detected. Regarding the spatial distribution of porpoises, the densities were generally higher in the western section of the study area. Hotspots were detected west and southwest of *alpha ventus*, at the Borkum Reef Ground (BRG) SCI and in areas nearby.

Density was estimated for 17 surveys (■ Fig. 13.8). The remaining two surveys did not cover the area appropriately. The lowest porpoise densities were estimated in 2009. In the following years, the overall density increased and pronounced seasonal differences were detected. The 2010 surveys show good temporal coverage between March and October, with lower densities in spring and autumn, and the highest in summer. This seasonal pattern repeated over the years.

A trend analysis based on aerial surveys conducted in the southern German North Sea revealed a significant increase in porpoise occurrence between 2002 and 2012 (Peschko et al. unpublished data). This trend was more pronounced in the western than the eastern section of the study area. Moreover, porpoise encounter rates were found to be significantly higher in the west, an area which includes the BRG SCI. An increasing trend in porpoise occurrence has been observed since 2005, i. e. before construction and operation of *alpha ventus* began. It is very unlikely that this small-scale windfarm has an impact on densities over the large monitored area. The recent increase in porpoise occurrence is most likely driven by natural variation, such as improved prey availability throughout the southern North Sea and/or is caused by reduced habitat suitability in the northern North Sea, as SCANSII reported a marked shift in distribution from north to south (Hammond et al. 2013).

It must be noted, however, that the BARD Offshore 1 and Trianel Windpark Borkum offshore windfarms were also constructed during the period 2010 to 2012 and that these and the BorWin alpha

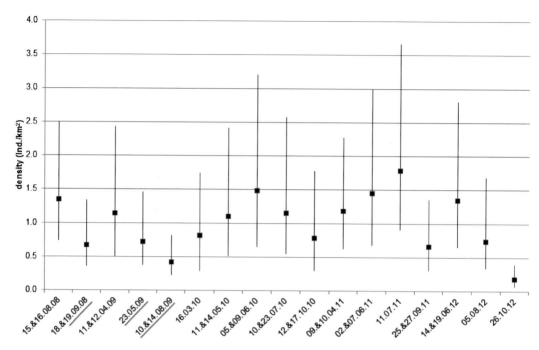

■ **Fig. 13.8** Estimated density of harbour porpoises in the study area. Error bars show 95 % confidence limits. Surveys conducted during pile driving or within 48 hours following pile driving at *alpha ventus* are underlined in red.

converter platform, installed in 2009, may have had an impact on the results.

13.3.3 Effects from the wind turbines during operation

Modelling was conducted using GAMLSS (Rigby & Stasinopoulos 2005) – a library for the open source statistics software R (R Development Core Team 2012). The data for 2011 (dph from click detectors deployed in the same design as for the construction period) was split and one model was constructed using a 'zero inflated binomial' distribution with data from periods when the turbines were switched off. The dph (transformed for presence/absence only) were used as the dependent variable, and the independent variables were distance to the windfarm, wind speed and the number of registered clicks (in four classes from 0 to 50,000 clicks/h; data were truncated at 50,000 clicks/h before analysis). Data was sparse for times when wind turbines were switched off. Hence, a random sample was selected from the data set for operational wind turbines to

provide the same sample size as for the 'off' situation.

The resulting dataset is small (310 samples) and hence the analysis shows wide confidence limits (■ Fig. 13.9). Especially for effects close to the windfarm, the analysis does not show a clear result due to large modelled variation.

It must, however, be noted that the closest C-POD position was deployed at 0.5 km from the nearest turbine. It is unlikely that an artificial reef effect or a noise effect would have significant impact on the registration rates at that station. The analysis hence shows that while very close range effects may occur, they most probably do not have a larger scale effect on porpoise detection rates.

13.4 Discussion

During the course of the study at the *alpha ventus* test site, more results became available from other windfarms, as did additional results from *alpha ventus* by means of the StUK3 monitoring programme. The StUK3 results indicate, similarly to the findings

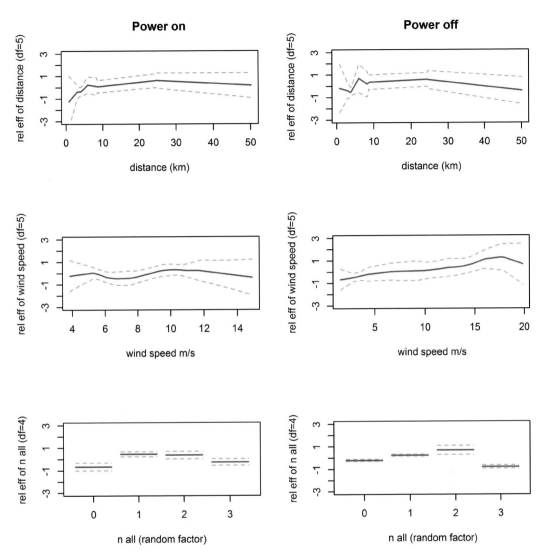

□ Fig. 13.9 Two GAMM model of the data from 2011, wind turbines are on (left) and off (right), nall = number of all recorded clicks in classes of 10,000, distance in km, wind speed in m s⁻¹, N = 310. Intercept of the model and distance to the windfarm significantly influence detection rates and the factorial variable 'all registered clicks' (n all). There is no clear trend in the distance and confidence limits are wide for data close to the windfarm, indicating that the data are not robust enough to show an effect from operation at distances close to the windfarm.

in SAM, that a strong seasonal pattern can be seen in acoustic monitoring, with the highest acoustic detection rates in March/April, the lowest in May to July, and a rise again towards the end of the year (Hansen et al. 2013). Furthermore, detection rates around the BRG SCI do not show obvious seasonal patterns but are higher overall than around *alpha ventus*. Seals have been regularly observed within the surveyed area (Hansen et al. 2013).

If porpoises stay within the general area during pile driving events as documented here, they may be subject to multiple displacements. Very short piling events and hence multiple successive exposures around the level where a temporary hearing impairment may happen, but above the behavioural reaction threshold, may lead to an accumulation of sound impact on the hearing system of porpoises. These events can elicit a temporary, and if sustained, a permanent hearing impairment due to a cumulative effect. This is in contrast to keeping the pile driving times as short as possible to minimise the temporal displacement effect and needs to be discussed further in the future.

Results from the Dutch Egmond aan Zee offshore windfarm show that detection rates in the windfarm can increase during the operation period in comparison to the pre-construction period. This is most probably linked to the windfarm being a sheltered area with no fishing effort and/or due to the artificial reef effect (Scheidat et al. 2011). In contrast, a probable longer-lasting effect of the construction period was observed at Denmark's Nysted offshore windfarm in the Baltic Sea, with a possible very slow recovery of porpoise occurrence from the construction period in 2002/03. No complete recovery due to habituation or an artificial reef effect was documented until the operational period in 2011/2012 (Teilmann & Carstensen 2012) – even though only limited pile driving was conducted for this windfarm, which has gravity-base foundations. For another windfarm in the Dutch North Sea, the Prinses Amalia Offshore Windpark, initial assessment does not point towards an increase in detection rates after the windfarm was built (van Polanen Petel et al. 2012).

Monitoring of pile driving and SAM at Horns Rev 1 (Tougaard et al. 2009a) and Horns Rev 2 (Brandt et al. 2011) point towards an avoidance radius of around 20 km during installation of monopiles. These results compare well with the displacement distances detected during monitoring at *alpha ventus*.

13.5 Perspectives

A large number of offshore windfarms will be built in the decades ahead. In Germany, the currently discussed target is the installation of 25 GW rated power by 2030, which would result in approximately 300 turbines being erected each year in the North and Baltic Sea over a 17-year period (assuming that each turbine has a rated power of 5 MW). This period is long in comparison to the life span of porpoises and seals and cumulative effects therefore need to be considered. The same applies to the effects of several windfarms erected at the same time

due to possible multiple displacements of marine mammals between piling sites. But cumulative impacts may also occur because of other anthropogenic stressors such as pollution, by-catch in fisheries, prey depletion and other noise impacts like underwater explosions or geophysical surveys. In future, these impacts must be seen in a larger framework to predict the biological significance of a single impact and to determine which stressor requires most management. This calls for extensive long-term monitoring and the StUK4 (BSH 2013) will play a major role in conjunction with other monitoring programmes in this regard. Results of pathological investigations and other research programmes (e.g. acoustic and satellite telemetry) must also be considered.

Whether offshore windfarm construction, operation and decommissioning have an impact on population level for marine mammals, remains unclear. The outcome may depend on the following factors:

- Is pile driving conducted in a sensitive area, where a large proportion of the local population occurs and during times when mammals are sensitive to impacts? For instance during birth and weaning period?
- How well can noise mitigation be incorporated into the construction process and can alternative, less noisy installation techniques be used instead of pile driving?
- During the operation period prey distribution may be altered due to artificial reef effects and may have an effect on marine mammal distribution as well. If prey species composition is altered, how does it affect marine mammals?

Furthermore, the effect of the construction of windfarms will strongly depend on (sub-) population status and dynamics – a probably healthy population of harbour porpoise, like in the North Sea, must be treated differently than the Baltic Sea population with numbers in the low hundreds (Gallus et al. 2012).

13.6 Acknowledgments

Thanks go to K. Krügel, A. Brandecker and A. Herrmann for C-POD calibration, to N. Tregenza, and

J. and J. Loveridge for support for C-PODs, to the aerial survey team consisting of S. Billerbeck, H. Feindt-Herr, T. Kesselring, l. Lehnert, D. Martensen-Stagginus, S. Müller, C. Rocholl and C. Schmidt. Thanks also for safe passages with Sylt Air, L. Petersen, FLM Aviation, KfK Venus and Skibsprojekt Christoffer ApS and FOGA Consult Aps. S. Paul, A. Hock, L. Weirup, A. van Neer, J. Sundermeyer, A. Ruser, S. Viquerat and B. Unger were involved in the data analysis. L. Kolbe, H. Boyens, P. Stührk and O. Meyer-Klaeden are thanked for C-POD servicing. Grateful thanks also go to project administrator S. Carstens-Michaelis. The authors would also like to acknowledge the help and ongoing support received from Federal Maritime and Hydrographic Agency (BSH), and Projektträger Jülich (PTJ).

Noise mitigation requirements (Incidental Provision 14)

Anika Beiersdorf

Without noise mitigation, it cannot be ruled out that pile driving will have a significant impact on marine mammals. The Federal Maritime and Hydrographic Agency (BSH), as the licensing authority for offshore installations in the German Exclusive Economic Zone (EEZ), therefore only permits pile driving if windfarm operators apply adequate noise mitigation to minimise sound emissions from noise-intensive construction work (installing foundations). Incidental Provision 14 in offshore windfarm approvals granted by BSH makes noise mitigation a condition of approval and sets out requirements for noise mitigation measures. It also includes instructions for measuring waterborne noise.

- A noise mitigation concept must ensure that the sound exposure level (SEL) does not exceed 160 dB re 1 µPa^{2}s and the peak level (L$_{peak}$) does not exceed 190 dB re 1 µPa outside of a 750 m radius of the construction site. During foundation work and wind turbine erection, the most suitable noise minimisation method (see Chap. 16) must be applied, taking into account both the best available technology as well as project and site-specific conditions.

- The noise mitigation concept must include a detailed description of the proposed mitigation methods and how mitigation procedures are implemented in the construction process, and must attest to the offshore suitability of the methods used.

- The windfarm operators must compile a noise emission forecast reporting on background noise in the proposed windfarm area and expected noise emissions during the construction work.

- The noise mitigation concept and forecast must be submitted to BSH for validation and assessment of potential environmental impacts at least 12 months before construction of the offshore windfarm.

- During noise-intensive construction work (impact pile driving), underwater noise must be monitored at various distances. The effectiveness of protective measures and noise mitigation measures must also be verified using underwater sound measurements.

- Suitable deterrent devices (such as pingers and seal scarers) must be used before pile driving starts to prevent marine mammals in the immediate vicinity of the piling site from being injured or even killed during noise-intensive construction works. Pile driving must begin at low energy ('soft start') and increase gradually.

- Noise emissions and the effectiveness of noise mitigation and deterrence measures must be documented and reported to BSH at short notice (within 24 hours).

- All construction work and the implementation of noise mitigation measures are closely supervised by BSH. In particular, BSH reserves the right to require additional noise mitigation measures.

Literature

Betke K & Matuschek R (2011). Messungen von Unterwasserschall beim Bau der Windenergieanlagen im Offshore-Testfeld *alpha ventus*. Abschlussbericht zum Monitoring nach StUK3 in der Bauphase, ITAP, Oldenburg, 48 pp.

Brandt MJ, Diederichs A, Betke K, Nehls G (2011). Responses of harbour porpoises to pile driving at the Horns Rev II offshore windfarm in the Danish North Sea. Marine Ecology Progress Series 421:205–216. doi: 10.3354/meps08888

Brandt MJ, Höschle C, Diederichs A, et al. (2013). Seal scarers as a tool to deter harbour porpoises from offshore construction sites. Marine Ecology Progress Series 475:291–302. doi: 10.3354/meps10100

Brandt MJ, Höschle C, Diederichs A, et al. (2012). Far-reaching effects of a seal scarer on harbour porpoises, *Phocoena phocoena*. Aquatic Conservation: Marine and Freshwater Ecosystems 23:222–232. doi: 10.1002/aqc.2311

BSH (2007). Standard Investigations of the impacts of offshore wind turbines on the marine environment (StUK3). Bundesamt für Seeschifffahrt und Hydrographie, Hamburg and Rostock, 58 pp.

BSH (2013). Standard Investigations of the impacts of offshore wind turbines on the marine environment (StUK4). Bundesamt für Seeschifffahrt und Hydrographie, Hamburg and Rostock, 83 pp.

Dähne M, Gilles A, Lucke K, et al. (2013a). Effects of pile-driving on harbour porpoises (*Phocoena phocoena*) at the first offshore windfarm in Germany. Environmental Research Letters 8:025002 (16pp). doi: 10.1088/1748-9326/8/2/025002

Dähne M, Verfuß UK, Brandecker A, et al. (2013b). Methodology and results of calibration of tonal click detectors for small odontocetes (C-PODs). Journal of the Acoustical Society of America 134:2514–2522. doi: 10.1121/1.4816578

Derweduen J, Vandendriessche S, Willems T, Hostens K (2012). The diet of demersal and semi-pelagic fish in the Thorntonbank windfarm: tracing changes using stomach analyses data. In: Degraer S, Brabant R, Rumes B (eds) Offshore windfarms in the Belgian part of the North Sea: Heading for an understanding of environmental impacts. Royal Belgian Institute of Natural Sciences, Management Unit of the North Sea Mathematical Models, Marine ecosystem management unit., pp 73–84.

Diederichs A, Hennig V, Nehls G (2008). Investigations of the bird collision risk and the responses of harbour porpoises in the offshore windfarms Horns Rev, North Sea, and Nysted, Baltic Sea, in Denmark Part II: Harbour porpoises (FKZ 0329963 + FKZ 0329963 A), Final Report. University Hamburg & BioConsult SH, Hamburg and Husum, 100 pp.

Gallus A, Dähne M, Verfuß UK, et al. (2012). Use of static passive acoustic monitoring to assess the status of the "Critically Endangered" Baltic harbour porpoise in German waters. Endangered Species Research 18:265–278. doi: 10.3354/esr00448

Gilles A, Adler S, Kaschner K, et al. (2011). Modelling harbour porpoise seasonal density as a function of the German Bight environment: implications for management. Endangered Species Research 14:157–169. doi: 10.3354/esr00344

Gilles A, Scheidat M, Siebert U (2009). Seasonal distribution of harbour porpoises and possible interference of offshore windfarms in the German North Sea. Marine Ecology Progress Series 383:295–307. doi: 10.3354/meps08020.

Hammond PS, Berggren P, Benke H, et al. (2002) Abundance of harbour porpoise and other cetaceans in the North Sea and adjacent waters. Journal of Applied Ecology 39:361–376.

Hammond PS, Macleod K, Berggren P, et al. (2013). Cetacean abundance and distribution in European Atlantic shelf waters to inform conservation and management. Biological Conservation 164:107–122. doi: 10.1016/j.biocon.2013.04.010.

Hansen S, Höschle C, Diederichs A, et al. (2013). Offshore-Testfeld *alpha ventus* Fachgutachten Meeressäugetiere 2. Untersuchungsjahr der Betriebsphase (Januar–Dezember 2011). IfAÖ und BioConsult SH, Husum, 87 pp.

Herr H, Scheidat M, Lehnert K, Siebert U (2009). Seals at sea: modelling seal distribution in the German bight based on aerial survey data. Marine Biology 156:811–820. doi: 10.1007/s00227-008-1105-x.

Kyhn LA, Tougaard J, Thomas L, et al. (2012). From echolocation clicks to animal density – Acoustic sampling of harbor porpoises with static dataloggers. Journal of the Acoustical Society of America 131:550–560. doi: 10.1121/1.3662070.

Lucke K, Siebert U, Lepper P a, Blanchet M-A (2009). Temporary shift in masked hearing thresholds in a harbor porpoise (*Phocoena phocoena*) after exposure to seismic airgun stimuli. Journal of the Acoustical Society of America 125:4060–70. doi: 10.1121/1.3117443-

Madsen P, Wahlberg M, Tougaard J, et al. (2006). Wind turbine underwater noise and marine mammals: implications of current knowledge and data needs. Marine Ecology Progress Series 309:279–295.

Van Polanen Petel T, Geelhoed S, Meesters E (2012). Harbour porpoise occurrence in relation to the Prinses Amaliawindpark. Report Number C177/10, Imares, Wageningen, 34 pp.

R Development Core Team (2012). R: A language and environment for statistical computing.

Rigby RA & Stasinopoulos DM (2005). Generalized additive models for location, scale and shape,(with discussion). Applied statistics 54:507–554.

Scheidat M, Tougaard J, Brasseur S, et al. (2011). Harbour porpoises (*Phocoena phocoena*) and windfarms: a case study in the Dutch North Sea. Environmental Research Letters 6:025102. doi: 10.1088/1748-9326/6/2/025102.

Scheidat M, Verdaat H, Aarts G (2012). Using aerial surveys to estimate density and distribution of harbour porpoises in Dutch waters. Journal of Sea Research 69:1–7. doi: 10.1016/j.seares. 2011.12.004.

Teilmann J & Carstensen J (2012). Negative long-term effects on harbour porpoises from a large scale offshore windfarm in the Baltic – Evidence of slow recovery. Environmental Research Letters 7:045101 (10 pp). doi: 10.1088/1748-9326/7/4/045101.

Thomas L, Buckland ST, Rexstad EA, et al. (2010.) Distance software: design and analysis of distance sampling surveys for estimating population size. The Journal of Applied Ecology 47:5–14. doi: 10.1111/j.1365-2664.2009.01737.x.

Tougaard J, Carstensen J, Teilmann J, et al. (2009a). Pile driving zone of responsiveness extends beyond 20 km for harbor porpoises (*Phocoena phocoena* (L.)). Journal of the Acoustical Society of America 126:11–14. doi: 10.1121/1.3132523.

Tougaard J, Henriksen O, Miller L (2009b). Underwater noise from three types of offshore wind turbines: estimation of impact zones for harbor porpoises and harbor seals. Journal of the Acoustical Society of America 125:3766–3773. doi: 10.1121/1.3117444.

TSEG (Trilateral Seal Expert Group) (2013). Aerial surveys of Harbour Seals in the Wadden Sea in 2013. Trilateral Seal Expert Group, Wilhelmshaven 3 pp.

Verfuß UK, Dähne M, Gallus A, et al. (2013). Determining the detection thresholds for harbor porpoise clicks of autonomous data loggers, the Timing Porpoise Detectors. Journal of the Acoustical Society of America 134:2462–2468. doi: 10.1121/1.4816571.

Marine habitat modelling for harbour porpoises in the German Bight

Henrik Skov, Stefan Heinänen, Dennis Arreborg Hansen, Florian Ladage,
Bastian Schlenz, Ramūnas Žydelis, Frank Thomsen

Federal Maritime and Hydrographic Agency,
Federal Ministry for the Environment, Nature Conservation and Nuclear Safety (Eds.)
Ecological Research at the Offshore Windfarm alpha ventus,
DOI 10.1007/978-3-658-02462-8_14, © Springer Fachmedien Wiesbaden 2014

14.1   Introduction

In the past decade, research in line with Federal Maritime and Hydrographic Agency (BSH) standards has yielded large amounts of data on the distribution and abundance of the harbour porpoise, *Phocoena phocoena*, in German waters (BSH 2007; see for example Scheidat et al. 2004, Thomsen et al. 2006). The StUKplus research project added further valuable information on the ecology of the species. Yet the data remains patchy, fragmented and often uneven, leaving many uncertainties, and a unified understanding of harbour porpoise distribution in German waters is still beyond reach. The environmental factors governing movements of the species in German waters, for example, are yet to be conclusively identified. Despite areas proposed as important for harbour porpoises in the German Bight, this makes it hard to define what constitutes 'good' porpoise habitats. Such information is needed as a key first step in any future marine spatial planning in the German Exclusive Economic Zone (EEZ). Given the highly ambitious plans for offshore windfarms in the German Bight, sound predictions of the occurrence of key species such as harbour porpoise are all the more vital in order to manage any conflicts between nature and human users. The harbour porpoise habitat modelling project aims to synthesise the data by long-term analysis and modelling of harbour porpoise distribution in the German Bight, i. e. the *alpha ventus* area and adjacent waters.

The study uses predictive distribution models to estimate harbour porpoise distributions in the German Bight during the summer season (June to August). The statistical models link observations from aerial surveys to oceanographic conditions and pressures. Oceanographic conditions during porpoise surveys were extracted from hydrodynamic models capable of describing patterns and dynamics of feeding habitats. As the modelling depends on the availability of coherent oceanographic and pressure data for the entire German Bight, the approach used combines hydrodynamic and noise models designed for the purpose. ◻ Figure 14.1 gives an overview of the marine habitat modelling approach used in the porpoise modelling project.

14.2   Methods

14.2.1   Survey data

Aerial line transect data was collated from the various offshore windfarm EIA baseline surveys carried out in the German EEZ and from research on the *alpha ventus* test site between 2002 and 2012. Initial screening removed biased data as follows:

- Counts with no positions or erroneous positions
- Counts done in sea states above Beaufort 2

The outcome was a large database of results from dedicated porpoise surveys at altitudes of 183 m (600 ft) and seabird surveys at altitudes of 76 m (250 ft) undertaken by a variety of institutions. ◻ Figure 14.2 gives an overview of the survey effort and the sightings obtained. More detailed analysis of the data is being undertaken, and the results of this analysis will be presented at a later date. In this chapter, initial results are presented based on large-scale distribution patterns. Models were first developed using dedicated porpoise and seabird surveys combined and then using dedicated porpoise surveys taken in isolation. As both approaches showed similar results, the combined model was used to take advantage of the larger number of samples.

14.2.2   Habitat modelling

The environmental variables used as predictors in the harbour porpoise habitat model were taken from hydrodynamic simulations both with and without post-processing (see details below), from topographic and geological data layers made available by BSH, and from underwater noise modelled from ship density (AIS) data. The core habitat variables are those reflecting flow characteristics of the German EEZ as set out below (current velocity, vertical velocity, frontal strength, and eddy activity). Water depth, sediment grain size, the proportions of fine, medium and coarse sand and gravel, and distance to gravel were included as static habitat variables.

Temperature and salinity were included to reflect the general characteristics of the water masses

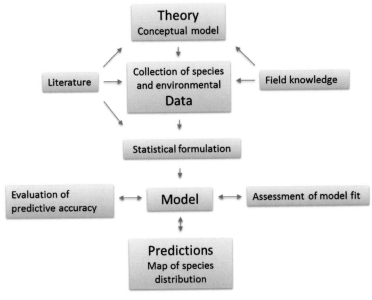

in the region. Two pressure variables – distance to land and underwater noise from ships – were included in the habitat models to account for potentially large adverse impacts of human infrastructures and activities on marine mammals (see OSPAR 2009 for a comprehensive review). The developed predictor variables are listed in 🔲 Table 14.1.

The relationships between the harbour porpoise survey data and the environmental features described by the habitat and pressure variables were analysed by using the multivariate statistical technique of generalised additive modelling (GAM). An overview of the habitat modelling workflow is shown in 🔲 Fig. 14.3. To describe and model the 'persistency' of harbour porpoise occurrences in the survey data a persistency index was calculated: the percentage occurrence in each 5×5 km grid cell (number of presences/conducted surveys) weighted (multiplied) by the standardised numbers of porpoises encountered per occurrence (standardised to the range 0–1). To avoid undersampled cells, all cells with a survey number less than 3 were excluded (🔲 Fig. 14.4). The persistency index was included as the response variable in the GAM model. As the response is proportional (between 0 and 100 %), a quasi-binomial model was used.

All variables that were not too closely correlated were fed into a first full model, from which unimportant variables were then excluded (Wood & Augustin 2002). The number of surveys carried out in each cell was used as a weighting in the GAM formula to offset survey effort bias. To assess predictive accuracy, the model was fitted on 70 % of the data (calibration data set) and evaluated on the remaining 30 % (evaluation data set). The model was assessed using Pearson's and Spearman's rank correlations (Potts & Elith 2006). The predictions were also mapped against the observations to verify that the patterns were reliable.

14.2.3 Hydrodynamic model

This section presents the hydrodynamic modelling methodology used to develop oceanographic variables for habitat modelling. Water level and current data were taken from DHI's hydrodynamic database for the North Atlantic (🔲 Fig. 14.5) enhanced by adding oceanographic measurements to the database. The regional North Atlantic model provided boundary data for the refined and more local German Bight model using DHI's MIKE 21 Flow Model FM (DHI 2009, 2013). The model complex uses a flexible mesh based on unstructured triangular or quadrangular elements. The entire model domain is shown in 🔲 Fig. 14.5. Data assimilation was applied at 17 stations to improve the perfor-

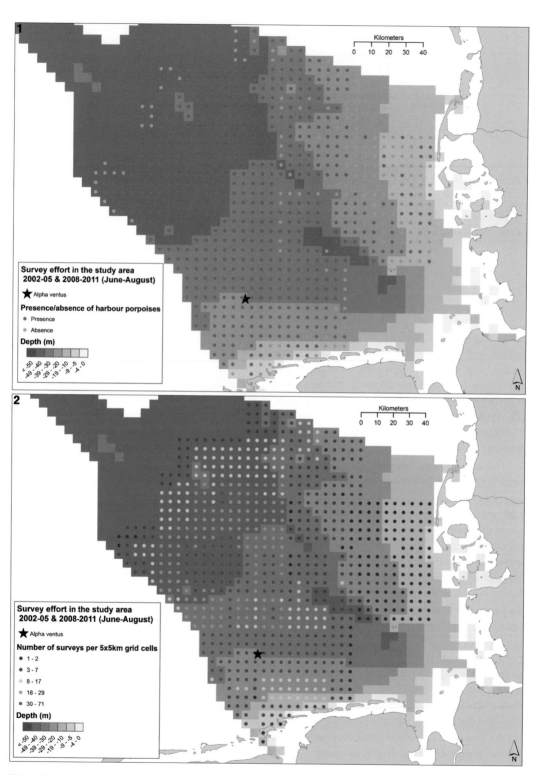

□ **Fig. 14.2** Aerial survey data collected during the summer period (June to August) 2002–2012. (1) The upper map shows the presence/absence of harbour porpoises and (2) the lower map the number of surveys conducted in each cell.

◧ **Table 14.1** List of predictor variables and their potential use.

Parameter Name	Parameter Type	Parameter Behaviour	Description	Source
Depth	Static habitat variable	Static	Water depth (cm)	BSH 500 m bathymetry
Slope	Static habitat variable	Static	Bottom slope (degrees)	Based on depth
Surface substrate – fine, medium, coarse sand and gravel	Static habitat variable	Static	Proportion (%) of substrate class	BSH
Surface substrate – Median grain size	Static habitat variable	Static	Median grain size (mm)	BSH
Distance to gravel > 5 %	Static habitat variable	Static	Euclidian distance to areas with proportion of gravel > 5 %	Based on % gravel (BSH)
Distance to land	Pressure variable	Static	Euclidian distance (m) to shore (East and North Frisian Islands) included	DHI
Underwater noise due to shipping	Pressure variable	Static (average)	Received sound pressure level (dB)	DHI/Danish Maritime Authority (AIS data)
Frontal strength	Dynamic habitat variable	Time-varying (average)	Local horizontal gradient of currents (m/s/m)	DHI – local hydrodynamic model (MIKE 21)
Salinity (average and standard deviation)	General water mass characteristic	Time-varying (average & SD)	Salinity (PSU) at 7 m depth	DHI – North Sea hydrodynamic model (MIKE 3)
Temperature (average and standard deviation)	General water mass characteristic	Time-varying (average & SD)	Water temperature (°C) at 7 m depth	DHI – North Sea hydrodynamic model (MIKE 3)
Current velocity (average of U and V)	General characteristic	Time-varying (average)	Direction and strength in zonal (U) and meridional (V) velocity	DHI – local hydrodynamic model (MIKE 21)
Current speed	General characteristic	Time-varying (average)	Current speed (m/s)	DHI – local hydrodynamic model (MIKE 21)
Vorticity (average for all years)	Dynamic habitat variable	Time-varying (average)	Eddy activity measured as local vorticity (m^2 s) of flow	DHI – local hydrodynamic model (MIKE 21)
Vertical velocity (average for all years)	Dynamic habitat variable	Time-varying (average)	Local up-/downwelling measured as vertical velocity (m/s, positive for upward flow) at 7 m depth	DHI – North Sea hydrodynamic model (MIKE 3)

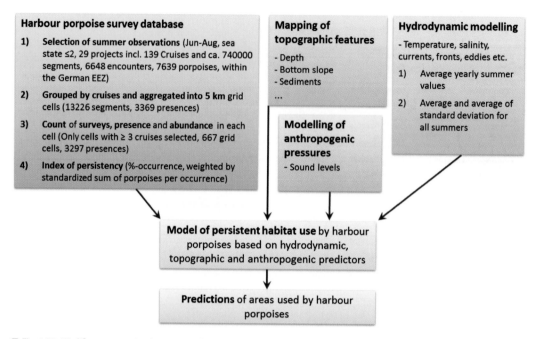

Harbour porpoise survey database

1) **Selection of summer observations** (Jun-Aug, sea state ≤2, 29 projects incl. 139 Cruises and ca. 740000 segments, 6648 encounters, 7639 porpoises, within the German EEZ)

2) **Grouped by cruises and aggregated into 5 km** grid cells (13226 segments, 3369 presences)

3) **Count of surveys, presence and abundance** in each cell (Only cells with ≥ 3 cruises selected, 667 grid cells, 3297 presences)

4) **Index of persistency** (%-occurrence, weighted by standardized sum of porpoises per occurrence)

Mapping of topographic features
- Depth
- Bottom slope
- Sediments
...

Hydrodynamic modelling
- Temperature, salinity, currents, fronts, eddies etc.
1) Average yearly summer values
2) Average and average of standard deviation for all summers

Modelling of anthropogenic pressures
- Sound levels

Model of persistent habitat use by harbour porpoises based on hydrodynamic, topographic and anthropogenic predictors

Predictions of areas used by harbour porpoises

◘ Fig. 14.3 Workflow, a stepwise description of the main working steps from data handling to predictions.

mance of the hydrodynamic hindcast model. The flow model was calibrated and validated for water level and currents against a number of stations in the North Sea, Danish waters and the Baltic Sea where tidal information or measurements were available.

The German Bight flow model has cell sizes in the highest resolution areas ranging from 0.2 km² in the inner regions of German Bight to approximately 3 km² in the outer region. The model domain for the German Bight is shown in ◘ Fig. 14.6. Water level and velocity components at model boundaries were extracted from the North Atlantic model. As with the North Atlantic model, the German Bight model was calibrated and validated for water level and currents against a number of stations.

The German Bight model data was validated against long-term observations for currents at FINO1 (2004–2008). As no reliable water level measurements are available for FINO1, water level observations from Heligoland were used (2004–2010). The Heligoland water level time series indicates good agreement of observed and modelled data (◘ Figs. 14.7, 14.8).

14.2.4 Modelling of underwater noise

This section presents the methodology for the acoustic modelling leading to the sound maps presented in the following and used as input for habitat modelling. The sound maps were based on information about bathymetry, water sound velocity and AIS shipping density maps. The bathymetry data is the same as in the hydrodynamic model. BSH provided sound speed profiles from the FINO1 research platform in the German Bight covering the four seasons with one sample during the winter and three samples each in spring, summer and autumn. ◘ Figure 14.9 presents the measured sound speed profiles. The profiles show little variation over depth, but variation over time.

Shipping density is based on AIS data provided by the Danish Maritime Authority for the period from 1 January to 31 March 2013. Although seasonal and annual variations in shipping density are likely in the German Bight, data from these three months is judged sufficient to retrieve a representative relative pattern of shipping density required for habitat modelling. The ship density map is shown in ◘ Fig. 14.10; it displays a range from just one ship passing to a few cells with more than 5,000 ships

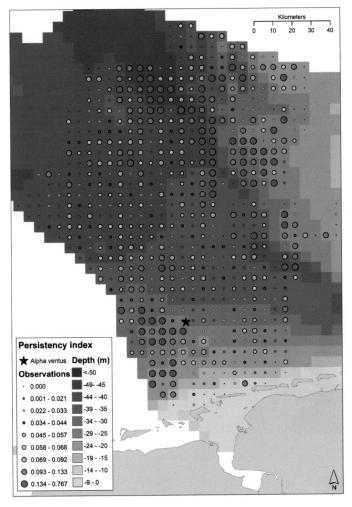

Persistency index

★ Alpha ventus   **Depth (m)**

Observations   ■ <-50

· 0.000   ■ -49- -45

· 0.001 - 0.021   ■ -44 - -40

∘ 0.022 - 0.033   ■ -39 - -35

∘ 0.034 - 0.044   ■ -34 - -30

○ 0.045 - 0.057   ■ -29 - -25

○ 0.058 - 0.068   ▨ -24 - -20

○ 0.069 - 0.092   ▦ -19 - -15

◐ 0.093 - 0.133   □ -14 - -10

● 0.134 - 0.767   □ -9 - 0

Kilometers
0   10   20   30   40

⬛ Fig. 14.4 The harbour porpoise data converted data converted into a persistency index: Percentage occurrence (occurrence/survey) multiplied by standardised abundance (range 0–1) per occurrence. Only part of the study area is shown for which sufficient survey effort was achieved.

passing during the three-month period. The number of ships was converted into a sound pressure level emitted from each cell, based on the number of ships in the cell and the sound pressure level emitted by a single ship. The latter value which was set at 175 dB re 1 µPa is an estimate of the mean sound pressure level per ship. In reality sound pressure source levels range from 160 dB re 1 µPa at 1 m for a fishing vessel to 204 dB re 1 µPa at 1 m for a supertanker (Ainslie et al. 2010). The resulting source SPL map is presented in ⬛ Fig. 14.11.

From the above sources, the sound propagation loss was estimated using a formula for transmission loss presented by Weston (1971), taking into account distance from source, water depth at source and receiver (porpoises), and strength of reflection

from the bottom. From each of the cells mentioned above, the received sound pressure level was calculated along 36 lines, spreading out from the centre of the cell to the edge of the domain.

The sound level was then interpolated onto a 1 × 1 km mesh, and finally summed over all source cells. In the resulting sound map, the sound value is the total, cumulated sound pressure level received at any given location during the three-month period (⬛ Fig. 14.12).

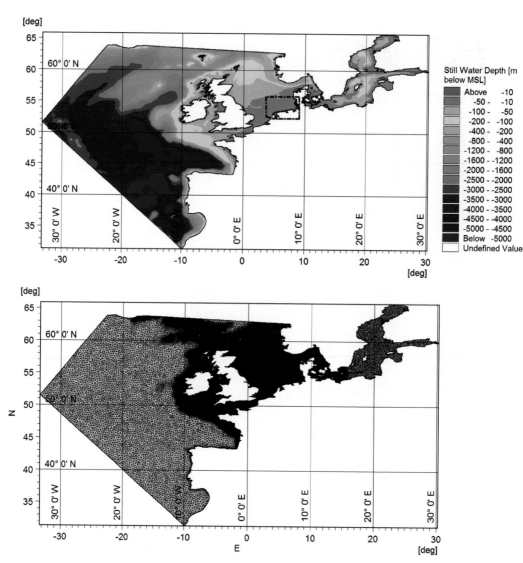

◘ Fig. 14.5 The North Atlantic model domain with bathymetry (top), grid of the regional North Atlantic model (bottom) and area of German Bight model (dashed box in top plot).

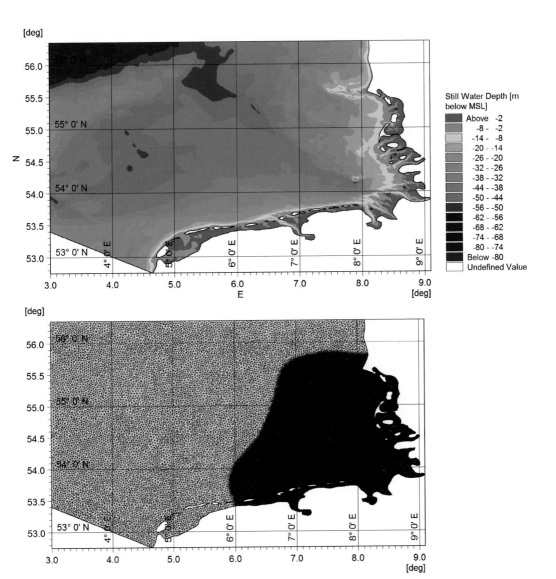

Fig. 14.6 The German Bight flow model domain with model grid.

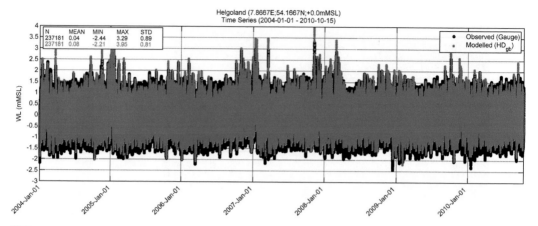

Fig. 14.7 Time series of observed and modelled water level (WL) at Heligoland.

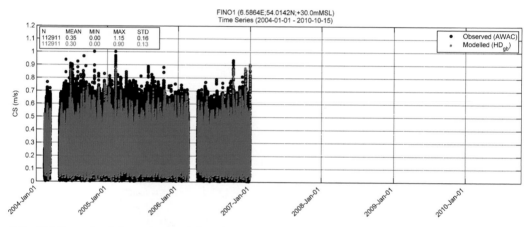

Fig. 14.8 Time series of observed and modelled current speed (CS) at FINO1.

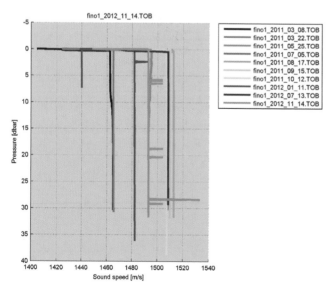

Fig. 14.9 Sound speed profiles (y-axis in [dbar] which indicate pressure and is equivalent to water depth).

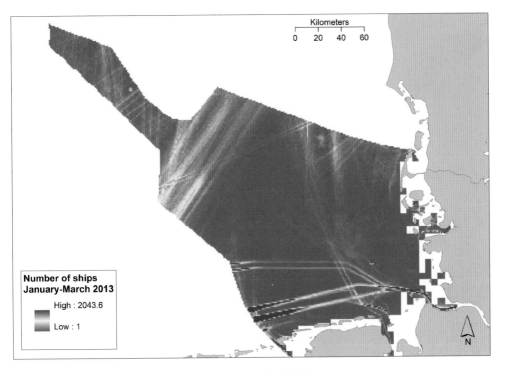

▫ Fig. 14.10 Number of ships in the German Bight, January–March 2013.

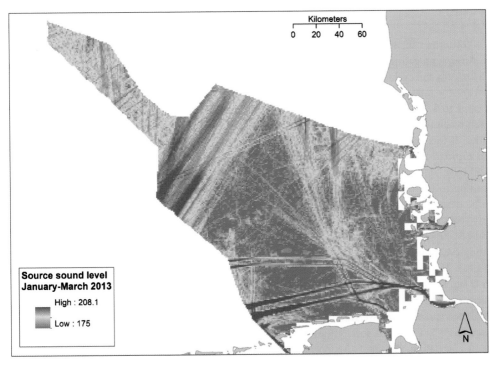

▫ Fig. 14.11 Source sound pressure level distribution [dB re 1 µPa @ 1 m], based on spatial analysis in cells.

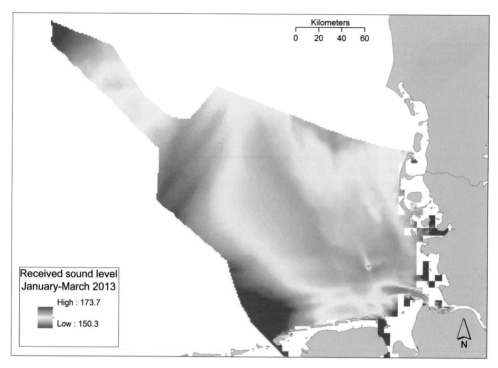

Fig. 14.12 Total, cumulated sound pressure level [dB 1 μPa] received.

14.3   Results

In total, 139 summer surveys were included in the analysis. After aggregating and merging the data into 5 x 5 km cells and excluding cells with less than three surveys, the total number of presences in the data set came to 3,297. By aggregating the data to a 5 × 5 km resolution and calculating the persistency index, it was possible to reduce the sample size from approximately 740,000 survey segments to 667 grid cells.

According to the model, the persistency pattern can be described as consisting of deep areas with fairly coarse surface sediments, intermediate north and east-flowing current velocities, relatively low sound levels and increasing current speed (◻ Fig. 14.13). The influential predictors are visualised in ◻ Fig. 14.14. The model explained 44.4 % of the variance in the data, and the correlation between predicted and observed weighted percentage occurrence was 0.59 (Pearson's correlation) or 0.49 (Spearman's correlation, ◻ Table 14.2). Correlation between predictors included in the model (which can invalidate predictions) was not a concern (< 0.7,

Dormann et al. 2012). Although some autocorrelation was present in the model residuals, the autocorrelation was low and the effect on the results is therefore small.

The predicted persistency of harbour porpoise summer occurrence in the German Bight is visualised in ◻ Figs. 14.15 and 14.16, including a classification of areas of different persistency classes defined by percentiles. The predictions indicate areas of high and very high persistency in three regions; one in the outer German Bight stretching northwest from Helgoland, one off the East Frisian islands and one just north of the Dogger Bank. The areas of high and very high persistency of harbour porpoises represent less than 20 % of the total area of the German EEZ. Areas of low persistency equalling 25 % of the German EEZ were predicted in the majority of the central parts of the EEZ deeper than 40 m, in the Elbe Valley, in the coastal zone (15 km wide) along the north Frisian coast and along the East Frisian islands (5 km wide), and on the southern part of Dogger Bank.

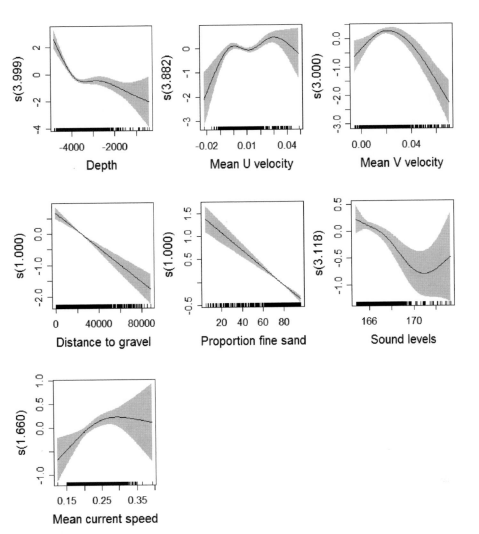

■ **Fig. 14.13** GAM response curves visualising the modelled responses between harbour porpoise persistency and the influential predictor variables included in the final model. The values of the predictors are displayed on the x-axis and the probability (for higher persistency) in logit scale (log(1/(1−x))) on the y-axis. The degree of smoothing is indicated in the title of the y-axis. The shaded areas show the 95 % confidence bands.

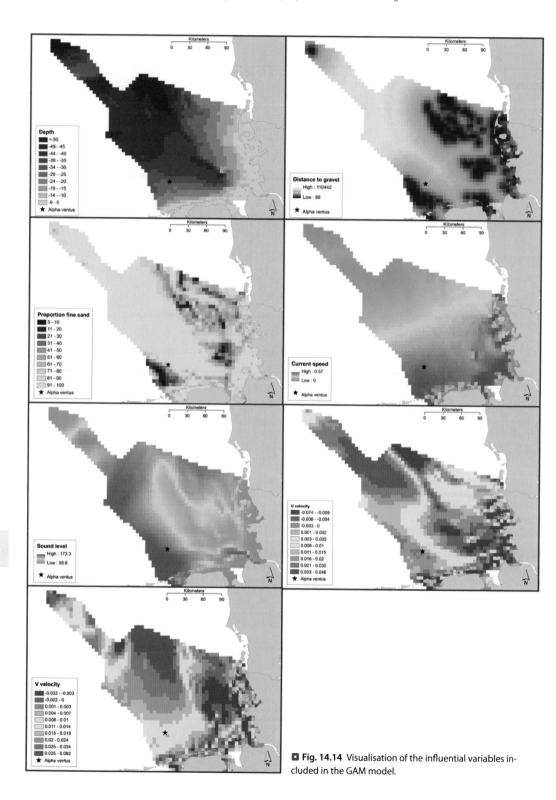

Fig. 14.14 Visualisation of the influential variables included in the GAM model.

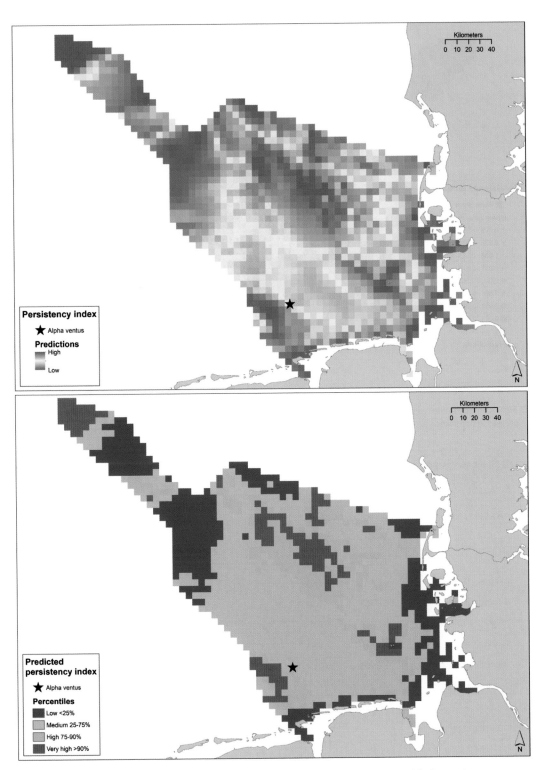

Fig. 14.15 Predicted persistency of harbour porpoise occurence in the German Bight, areas with higher persistency are indicated by a more reddish colour. The upper map is visualised with a continuous scale while the lower map is classified based on percentiles (25 %, 75 % and 90 %).

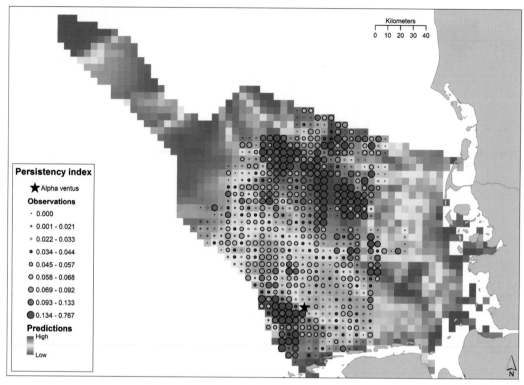

◩ Fig. 14.16 The predicted persistency of harbour porpoise occurrence overlaid with the observed persistency. Persistency is expressed as occurrence/survey multiplied by the standardised abundance (range 0–1) per occurrence.

◩ Table 14.2 Approximate significance (*p*-values) and F-values for the influential variables included in the model.

Variable	F	*p*-values
Depth	15.937	< 0.01
U velocity	5.727	< 0.01
V velocity	13.467	< 0.01
Distance to gravel > 5 %	54.327	< 0.01
Proportion fine sand	92.860	< 0.01
Sound levels	4.471	< 0.01
Mean current speed	3.596	0.03
Deviance explained	44.4 %	
N	667	
Evaluation results		
Pearson's correlation	0.59	
Spearman's correlation	0.49	

14.4 **Discussion**

The model results indicate that the summer distribution of harbour porpoises in the German Bight reflects flow patterns created by the specific hydrodynamic features of the region, which in turn could be an indication of prey distribution.

Three regions of persistent high presence of harbour porpoise were identified. One traverses the outer German Bight stretching from the shallows of Heligoland and extending in a northwesterly direction to the Danish border, largely covering the region between the 20 to 40 m depth contours. The eastern slopes of the Heligoland Channel are located centrally in this high presence region. The region of the eastern slopes of the Heligoland Channel coincides with the zone of the permanent dynamic salinity front (Becker & Prahm-Rodewald 1980, Dippner 1993), a zone characterised by high eddy activity, gradients in zonal and meridional current velocities and moderate mean frontal strength. From the salinity front and eastwards a belt of coarse sand and gravel is found. In terms of anthropogenic pressures, the salinity front is also characterised by moderate to low levels of underwater noise as it is located east of the main shipping lane.

Another region with high predicted presence is located just north of the East Frisian Islands (Borkum and Juist). This area is characterised by a high proportion of coarse and gravel sediments, and it coincides with the location of the boundary between stratified and well-mixed water masses where eddies and upwelling are produced in summer. Contrary to the situation in the northern German Bight, the level of underwater noise in this region is quite high due to proximity to the main shipping lane.

A third region of predicted high persistency of porpoise sightings is located on the northern flanks of the Dogger Bank – a region under strong influence from tidal currents creating regular intrusion of nutrient-rich bottom water and a permanent gradient between well-mixed water masses over the bank and stratified water masses to the north of the bank (Nielsen 1993, Pedersen 1994). The estimated noise levels in this region are quite low. It is important to note, however, that the predictions in this region are solely extrapolations from the relation-

ships within the survey effort. The same applies for the easternmost parts of the EEZ.

The three areas predicted by our model to be most persistently used by harbour porpoises correspond well with the high densities areas predicted by Gilles et al. (2011). However, our model indicates that the Borkum region in the south of the study area is also persistently used in summer, and not just in spring as shown by Gilles et al. (2011). The survey effort in this region is however much higher in comparison with the northeast region, and our dataset can therefore be said to be biased towards the south. To offset this bias, survey effort was used as a weighting in our GAM model. Interestingly, our model predictions extending outside the surveyed area also correspond well with the findings by Gilles et al. 2011, indicating that the variables included in the model are capable of predicting the areas most persistently used by porpoises in the German Bight.

The oceanographic and anthropogenic interpretations of the three identified areas with high predicted persistency of porpoise observations in summer fit well with the overall responses described by the model, i. e. higher persistency in deeper areas with a low proportion of fine sand and close to areas with gravel, moderate north and east-flowing current velocities and areas with decreasing sound levels. Having said that, it is obvious that each of the regions of predicted high persistency has characteristics that vary to a greater or lesser degree from the general responses picked up by the model. Such variation can be seen in the response to underwater noise, which stands out as a relatively important factor in harbour porpoise distribution overall, yet does not lead to lower predicted persistency in the somewhat noisy region off Borkum and Juist.

Similarly, because mean distributions are used, it is unlikely that a model with a 5 km resolution will have picked up temporal and spatial variations in porpoise distribution in response to dynamic habitat features like frontal, eddy and upwelling activity. To model harbour porpoise distribution in greater detail, the dynamic coupling would probably have to be analysed at the level of such features. Despite high explanatory power (44.4 % deviance explained) and predictive power (0.59 Pearson's correlation), the developed model thus has limitations when it comes to resolving fine-scale patterns in harbour porpoise distribution. Further work will attempt to improve the model by incorporating *in-situ* dynamic predictors at higher resolution based on a close spatio-temporal match between observations and the hydrodynamic model data.

Although the database of aerial surveys carried out by the various windfarm projects covers most of the modelled region of the German North Sea, the area north of the East Frisian Islands has been subject to relatively more effort than the rest of the area. To account for potential bias introduced by fewer surveys off the north Frisian islands, further work will attempt to resolve any year-to-year and seasonal variation in the responses to predictor variables and in the predicted distribution patterns.

14.5   Perspectives

The ambitious plans for sustainably harvesting energy from offshore windfarms in the German North Sea EEZ necessitate assessments of cumulative impacts both within the offshore wind sector and across all offshore sectors. Although the harbour porpoise model study is still in progress, the distribution models it produces are expected to allow prediction of seasonal harbour porpoise distributions across the German EEZ at fine spatial resolution for the 10-year investigation period. The resulting maps can be used to develop a decision support tool for verifying long-term predictions. Studies have shown that harbour porpoises show behavioural reactions to offshore windfarm construction using impact pile driving at distances of approximately 20 km from the source (e.g. Tougaard et al. 2009, Brandt et al. 2011). Based on these empirical studies, estimations of harbour porpoise habitat displacement due to the further development of offshore windfarms could be quantified. Estimations of habitat displacement need to consider the mitigation effect of noise mitigation measures and may be used to predict temporary and potentially prolonged distribution changes for a range of future development scenarios. Such a tool can be a major aid to spatial planning in licence applications for offshore windfarms and other human activities, by providing cumulative assessments of impacts due to multiple developments

in relation to the harbour porpoise population in the entire sector of the North Sea, and by making it possible to estimate the sustainability of each development scheme.

Literature

Ainslie MA (2010). Principles of sonar performance modelling, Vol. Springer in association with Praxis Publishing Chichester, UK.

Becker GA, Dick S, Dippner JW (1992). Hydrography of the German Bight. Mar. Ecol. Prog. Ser. 91: 9–18.

Becker GA, Prahm-Rodewald G (1980). Fronten im Meer. Salzgehaltsfronten in der Deutschen Bucht. Seewart 41: 12–21.

Brandt MJ, Diederichs A, Betke K, Nehls G (2011). Responses of harbour porpoises to pile driving at the Horns Rev II offshore windfarm in the Danish North Sea. Marine Ecology Progress Series 421:205–216. doi: 10.3354/meps08888.

BSH (2007). Standard Investigation of the Impacts of Offshore Wind Turbines on the Marine Environment (StUK3). Bundesamt für Seeschifffahrt und Hydrographie (BSH), Hamburg und Rostock, 58p.

DHI (2009). MIKE 21 FLOW MODEL FM – Hydrodynamic Module – User Guide.

DHI (2013). MIKE 21 & MIKE 3 Flow Model FM – Hydrodynamic and Transport Module – Scientific Documentation.

Dippner JW (1993). A frontal-resolving model for the German Bight. Cont Shelf Res 13: 49–66.

Gilles A, Adler S, Kaschner K, Scheidat M, Siebert U (2011). Modelling harbour porpoise seasonal density as a function of the German Bight environment. Endangered species research, 14: 157–169.

Nielsen TG, Løkkegaard B, Richardson K, Pedersen FB, Hansen L (1993). Structure of plankton communities in the Dogger Bank area (North Sea) during a stratified situation. Mar. Ecol. Prog. Ser. 95: 115–131.

OSPAR. (2009). Overview of the impacts of anthropogenic underwater sound in the marine environment: OSPAR Convention for the Protection of the Marine Environment of the North-East Atlantic.

Pedersen FB (1994). The Oceanographic and Biological Tidal Cycle Succession in Shallow Sea Fronts in the North Sea and the English Channel. Est. Coast.Shelf Science 38: 249–269.

Potts JM & Elith J (2006). Comparing species abundance models. Ecological Modelling, 199, 153–163.

Scheidat M, Kock KH, Siebert U (2004). Summer distribution of harbour porpoise (*Phocoena phocoena*) in the German North Sea and Baltic Sea. Journal of Cetacean Research and Management, 6, 251–257.

Thomsen F, Laczny M, Piper W (2006). A recovery of harbour porpoises (*Phocoena phocoena*) in the southern North Sea? A case study off Eastern Frisia, Germany Heligoland Marine Research, 60, 189–195.

Tougaard T, Carstensen J, Teilmann J, Skov H, Rasmussen P (2009). Pile driving zone of responsiveness extends beyond

20 km for harbor porpoises (*phocoena phocoena* (l.)). Journal of the Acoustical Society of America 126:11–14

Weston DE (1971). Intensity-range relation in oceanographic acoustics. Journal of Sound and Vibration 18(2), 217–287.

Wood SN & Augustin NH (2002). GAMs with integrated model selection using penalized regression splines and applications to environmental modelling. Ecological Modelling, 157, 157–177.

Underwater construction and operational noise at *alpha ventus*

Klaus Betke

Federal Maritime and Hydrographic Agency,
Federal Ministry for the Environment, Nature Conservation and Nuclear Safety (Eds.)
Ecological Research at the Offshore Windfarm alpha ventus,
DOI 10.1007/978-3-658-02462-8_15, © Springer Fachmedien Wiesbaden 2014

15.1 Introduction

15.1.1 Impact of underwater noise on marine mammals

Construction of offshore windfarms produces underwater noise that adversely affects the marine environment. Most offshore wind turbines are erected on driven pile foundations. In this technique, steel tubes with a diameter of several metres are driven up to 50 m into the seabed (Fig. 15.1). This is done with an impact pile driver (often simply called 'hammer') and causes strong, impulsive noise that can be harmful to animals close to the construction site. Even if the sound level is too low to cause physical damage to the auditory system, the noise can mask an animal's acoustic communication or affect the acoustic detection of prey. Harbour porpoises experience stress from noise, as can be seen from flight responses during pile driving operations (see Chap. 13).

Common marine mammals in the North Sea and the Baltic Sea are the harbour porpoise (*Phocoena phocoena*), the harbour seal (*Phoca vitulina*) and, less frequently, the grey seal (*Halichoerus grypus*). These species are protected by the EU's Habitats Directive. The harbour porpoise is also protected by the Agreement on the Conservation of Small Cetaceans of the Baltic, North East Atlantic, Irish and North Seas (ASCOBANS), which has been signed by ten European countries. In order to protect these species from unwanted effects of wind turbine construction noise, a limit has been adopted by law in Germany after evaluation of the noise measurements at *alpha ventus*: The sound exposure level of a pile driving strike must not exceed 160 dB re1 µPa² s outside of a circle of 750 m radius, and the peak level (L_{peak}) must not exceed 190 dB re1 µPa. As will be seen later, this dual threshold criterion is difficult to meet without noise mitigation measures.

Underwater noise is produced not only in the construction of offshore wind turbines, but also during their operation. Operating noise from onshore windfarms is dominated by flow noise from the rotor blades. This sound does not, however, play a role in underwater noise produced by offshore wind turbines, because the sound from the blades

is almost completely reflected at the sea surface and does not impinge into the water. A further component is noise from the generator and the gear box. Vibrations of these machinery parts are conducted downwards in the tower wall and are radiated as sound into the water. The sound is almost constant and tonal, i. e. it consists of distinct peaks in the frequency spectrum, and most of them can be associated with tooth mesh frequencies of the gear box. When listening to a signal picked up by a hydrophone near an offshore wind turbine, it sounds like a hum or a low musical note. It is much less intense than the construction noise, but on the other hand it is almost permanent.

15.1.2 Sound, sound levels and sound spectra

Sound is a disturbance in pressure that propagates through a medium like air or water. The sound pressure adds to the static or ambient pressure, as shown in Fig. 15.2. The ambient pressure is about 100,000 Pa at the water surface, 200,000 Pa at 10 m depth, and so on. In acoustics, it is usually disregarded. The second criterion for sound, propagation, is important because there are also pressure fluctuations that do not propagate. In water, these are caused by wave motions, currents or tide and may produce strong signals in a hydrophone. Sometimes it is difficult to keep this 'pseudo sound' low and to separate it from 'true' sound.

Sound pressure can vary over a large range. For this reason, it is common to specify sound on a logarithmic scale in terms of *levels* in decibels (dB), as it is known from communications engineering, rather than stating sound pressure numbers. There are various level quantities with different definitions. For this investigation, the following three are important:

- Equivalent continuous sound pressure level, L_{eq}
- Sound exposure level (or single event sound pressure level) SEL
- Peak level, L_{peak}

The **equivalent continuous sound pressure level** is an average level. The averaging time can be from

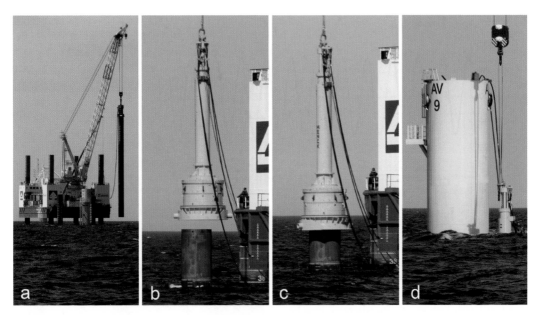

■ **Fig. 15.1** (a) Pile in full length, (b) Impact hammer, (c) Hammer in position, (d) Hammer working under water.

below one second to hours, whatever is appropriate for the problem. It is also called root mean square (RMS) level. The reason for this can be seen in the formula below:

$$L_{eq} = 20 \log_{10} \left(\frac{\sqrt{\overline{p^2}}}{p_0} \right)$$

First, the sound pressure p is squared, then the arithmetic mean of the p^2 values is computed (denoted by the bar above the p^2). The square root of this result is the RMS value of the sound pressure. Now the actual level in decibels is computed: The RMS value is divided by the reference pressure p_0 and the log function is applied to obtain the result in decibels. By international agreement, p_0 is 1 µPa. When dealing with decibels, the reference value should be stated. For the example in ■ Fig. 15.2, the calculation yields L_{eq} = 120.5 dB re 1 µPa.

The **sound exposure level** or SEL is calculated in a similar manner, except that the averaging time is one second, regardless of the duration of the sound. It is suitable to describe sound *events* rather than continuous sound. The SEL of a single strike has become a de facto standard for describing un-

derwater pile driving noise. The SEL is the level of a sound signal that is constant over one second and has the same acoustic energy as the pile strike. With one strike per second, SEL and L_{eq} are equal.

The **peak level** is not an average, but simply an indicator for the maximum sound pressure:

$$L_{peak} = 20 \log_{10} \left(\frac{p_{max}}{p_0} \right)$$

Here, p_{max} is the highest positive or highest negative sound pressure within an observation period. For pile driving noise, the L_{peak} value is typically 20 to 25 dB higher than the SEL.

So far only broadband levels have been considered, meaning that all portions of the sound along the frequency or 'pitch' axis are treated equally. But often it is useful or necessary to limit the frequency range, for example to exclude sound from an analysis that has nothing to do with the problem – like the low-frequency pseudo sound mentioned earlier. A useful representation of sound is the spectrum, where amplitude or intensity or level is plotted against a frequency axis. The curves in ■ Fig. 15.7, for example, are third-octave spectra: Each point covers a frequency band that corresponds to four adjacent piano

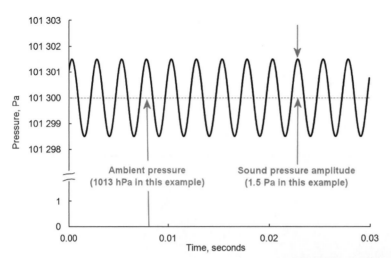

■ **Fig. 15.2** Sound pressure is a small disturbance of the hydrostatic pressure.

keys. This kind of representation is advantageous for hearing-related problems. It is to distinguish from a constant-bandwidth spectrum, where all bins have the same absolute width of 1 Hz, which is often more suitable for analysing technical processes.

15.2   Methods

15.2.1 Study design and layout: Construction phase

Between April and August 2009, an extensive monitoring programme was carried out at *alpha ventus* to determine and quantify the underwater noise produced by the installation of the wind turbine foundations. Two different designs were chosen as carrying structures: six jackets (AV1–AV6) and six tripods (AV7–AV12). During the installation of the tripod AV9, a prototype noise mitigation system was tested and evaluated.

All foundations are 'attached' to the sea floor with piles of 2.4 to 2.6 m diameter and a penetration depth of about 30 m. Hence, the pile driving technique used at *alpha ventus* was similar for all turbines, except that four piles are needed for the four-legged jackets and three piles for every three-legged tripod. The noise levels were also expected to be similar for all foundations. The number of hammer strikes that were necessary to install a pile varied between 2,400 and 8,700. The pile driving operations lasted usually two to three hours (net piling

■ **Fig. 15.3** Preparation of a sound recorder for deployment.

time). Due to breaks for adjustments, for moving the pile driver from one pile to the next or for picking up the next pile with the crane, the installation time for a complete foundation varied from a few hours to several days (Betke 2011).

The sound measurements took place between April and August 2009. Most recordings were made with autonomous systems (■ Fig. 15.3). These were deployed together with porpoise detectors (C-PODs) that were used for harbour porpoise monitoring during the construction phase (Diederichs et al. 2010). At each measurement position, the hydrophone was kept about 2 m above the ground by means of a float. The systems remained on site for 2 to 4 weeks, and the sound recordings were evaluated after recovery. The measurement positions are shown in ■ Fig. 15.4. Two to three

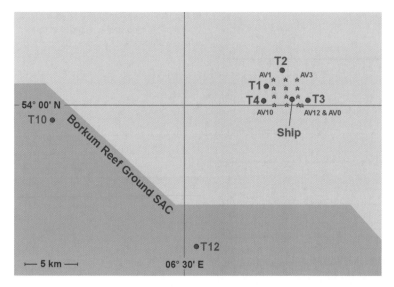

■ **Fig. 15.4** Sound measurement positions at porpoise detector locations T1, T2, etc. used by Diederichs et al. (2010). At AV5 and AV9, additional measurements were made by ship. AV0 is the transformer platform.

recording devices were active at the same time. Complete pile driving works were recorded from AV8 and AV9; parts were recorded from AV2, 3, 5, 6, 7, 10, 11 and 12. Measurement distances were between 800 and 2,400 m, except for positions T10 and T11 which were placed at about 16 km distance from *alpha ventus* within the Borkum Reef Ground SCI.

15.2.2 Study design and layout: Operation phase

Measurements were made at *alpha ventus* in May and June 2011. This was the first operating noise measurement on wind turbines in the 5 MW class; until then, only smaller turbines had been evaluated (Betke & Matuschek 2012). The equipment was almost the same as for recording construction noise, except that more sensitive hydrophones were used as the expected sound levels were much lower. Sound was recorded simultaneously at five points: 100 m from AV1 and from AV10, and at positions T3, T4 and T10 shown in ■ Fig. 15.4.

15.3 Results and discussion

15.3.1 Construction phase

■ Figure 15.5 shows the sound pressure of a typical pile driving strike at three different distances. Even at a relatively large distance of 17 km, the noise level is clearly above the ambient noise, although of course much weaker than close to the construction site. But also the shape changes due to propagation effects and reverberation in the sea. The impulse is prolonged with increasing distance from the source. Near the pile driver, it sounds 'dry', while at a few kilometres, it sounds more like a bass drum. An example for the sound level during the complete installation of all three piles of a foundation is given in ■ Fig. 15.6. The time from the first to the last pile strike was about 13 hours in this case. From the frequency spectra in ■ Fig. 15.7 it can be seen that the maximum frequency of the pile driving noise is in the range of 100 to 300 Hz which is typical for underwater pile driving noise.

The SEL in ■ Fig. 15.6 exceeds the noise threshold level of 160 dB re 1 μPa^2s for most of the time, even though the measurement distance was larger than the reference value of 750 m. Since the pile driving energy itself can hardly be reduced to make the hammering quieter, passive noise reduction is required. A special kind of noise barrier, a 'small bubble curtain', was developed and

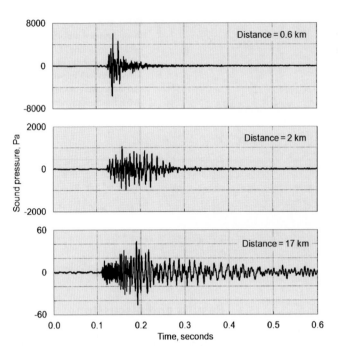

Fig. 15.5 Sound pressure versus time for a typical pile driving stroke, recorded at three different distances.

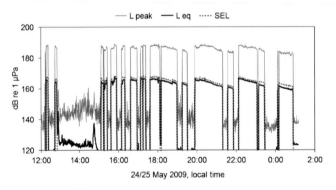

Fig. 15.6 Level versus time for the installation all three piles at AV8, recorded at 1,600 m distance.

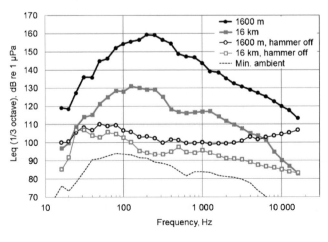

Fig. 15.7 Frequency spectrum of a pile driving strike at two different distances. The background level (dashed line) was measured in May 2008 before construction started.

□ Fig. 15.8 (a) Two of the three pile sleeves of the AV9 tripod were fitted with rings of perforated tubes in order to establish a bubble curtain. (b) The bubble curtain creates a ring of white water around the pile.

deployed. The prototype was tested during construction of the foundation of AV9 on 31 May 2009 (Rustemeier et al. 2012a). The principle is that bubbly water attenuates and reflects sound. A technical solution is to put perforated tubes around the noise source. Air is pressed into the tubes so that bubbles rise in the water and form a 'curtain' (□ Fig. 15.8). It was planned to extend the noise mitigation system that was attached to the pile sleeves of the AV9 foundation upward to the sea surface with a mobile stack of six additional rings of air tubes upward to the sea surface, but this experiment had to be abandoned because of impending bad weather.

Even with the incomplete setup of the small bubble curtain, the broadband SEL was reduced by about 12 dB (Rustemeier et al. 2012b). A problem with bubble curtains, however, is the drift of the bubbles due to tide current which may lead to 'holes' in the curtain that largely reduce its efficacy in the upstream direction. For this reason and for easier installation, a different system is often preferred, consisting of a perforated hose that is laid out on the sea floor at 70 to 100 m radius around the construction site. With the more recent installation of a 'big bubble curtain', similar noise reductions of 12 to 15 dB were achieved (Betke et al. 2012).

□ Figure 15.9 summarises the measurements at AV5, AV8 and AV9 and gives an idea of the level vari-

ation. For AV9, the spreading is larger because values with and without a bubble curtain are included.

15.3.2 Operation phase

During the course of the operating noise investigations at *alpha ventus,* it turned out that all measured operating noise could be assigned to the tripods AV7 to AV12. No noise recordings could be attributed to the jacket foundations. That may have to do with different generators and gear boxes, but also with the less massy design of the jacket structure of AV1 to AV6 that conducts machine vibration less efficiently into the water than tripods. A narrowband analysis revealed a prominent tone at a frequency of 90 Hz, with up to 120 dB re 1 µPa at 100 m distance from the tripod, and a few weaker ones between 400 and 800 Hz (□ Fig. 15.10). Depending on the operating condition of the turbine, the level of these tones varied by some ± 5 dB, and their frequencies varied by a few Hz. The levels are similar or lower compared to data measured in Danish offshore windfarms with turbines of the 2 MW class (Betke 2012). At the 15 km distant position T10, no operational noise was detected.

Above approximately 500 Hz, the tone level can exceed the hearing threshold of a harbour porpoise (Kasteleien et al. 2010), so that porpoises might be

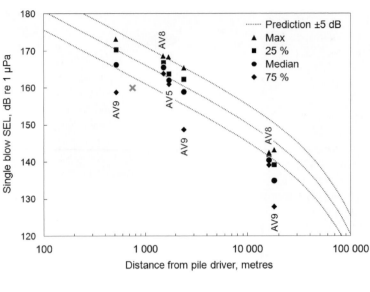

■ **Fig. 15.9** Sound exposure levels from pile driving works at AV5, AV8 and AV9 at various distances. The asterisk marks 160 dB at 750 m.

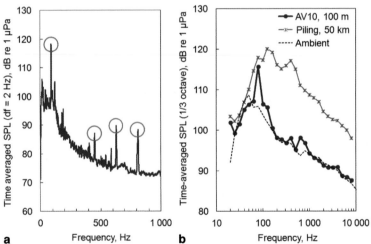

■ **Fig. 15.10** (**a**): Narrowband spectrum with tonal machinery noise components (*circled*) originating from AV10 running at its rated power of 5 MW. (**b**): In the 1/3 octave spectrum, all but the strong 90 Hz component vanish in the background level (the background curve reflects the same wind condition, but was recorded in 2008 before the turbines were installed). The curve labelled 'piling 50 km' is a pile driving noise spectrum from a distant offshore windfarm construction site observed on 2 June 2011 at the same position 100 m southwest of AV10.

able at least to hear some of the tones, although the perception alone will not necessarily induce a reaction. Furthermore, the hearing threshold is exceeded by the background noise. Above a certain frequency, this is virtually always the case, regardless of the presence of wind turbines. At present it is not known whether wind turbine operational noise has an impact on the behaviour of marine mammals.

15.4 Perspectives

It is generally agreed that keeping the SEL of a pile driving strike below 160 dB re 1 µPa²s outside of a circle of 750 m radius from the source is a safe value to guard marine mammals from physical injury. As a single-number value, it is also easy to handle for authorities. It should be kept in mind, however, that it is a 'technical' quantity. This is important at distances beyond the zone of possible physical injury, where the number may not be optimum to describe an animal's response to noise. This has several reasons: First, the number is an unweighted broadband level, with all frequency components of the sound treated equally. But an animal's auditory system is not equally sensitive to all frequencies. And noise mitigation techniques like bubble curtains do not reduce the level of all frequency components by the same amount; they alter the shape of the frequency spectrum. Hence if the efficacy of mitigation is described in terms of the broadband level, this may not adequately reflect the improvement for the animals. Secondly, unlike with the SEL, an animal's auditory system does not average the sound energy over one second to produce a 'loudness' sensation (actually the averaging time of a porpoise's hearing system is much shorter, in the order of 0.1 second). And finally, the installation of a pile needs more than one pile driving strike. That is, the parameter 'impact duration', which is not unlikely to affect the pressure on the marine environment, is not reflected in the current standard. Hence a future task should be to develop a noise descriptor that is more closely related to animals' perception of sound.

15.5 Acknowledgements

Special thanks to: R. Matuschek, B. Meenen, D. Galinski, A. Diederichs, G. Nehls, K. Blasche, T. Verfuß, C. Honnef, and the crews of R/V Tine Bødker and F/V Orion.

Measuring instruction for underwater sound monitoring

Anika Beiersdorf

As part of licensing procedures and during the construction and operation of offshore installations, it is essential to analyse the impact of underwater noise on the marine environment. The temporal and spatial scope of acoustic investigations is laid down in the BSH 'Standard for Environmental Impact Assessment' (StUK4, BSH 2013). This requires underwater sound measurements to be carried out prior to construction, during construction and during operation of an offshore windfarm.

BSH published a standardised guidance 'Offshore windfarms: Measuring instruction for underwater sound monitoring' in October 2011 (BSH 2011). Based on initial results from acoustic investigations at the *alpha ventus* test site (see above in this chapter) and offshore research platforms, the instruction gives detailed information on performing and analysing acoustic measurements. It became the legal basis for conducting investigations of underwater noise at offshore windfarms and supplements the StUK4. The instruction describes the general procedure for measuring underwater sound connected with the construction and operation of offshore windfarms. It covers the four licensing and enforcement phases for offshore installations in the German EEZ:

a) Baseline study/ preliminary investigations
b) Construction phase
c) Operation phase
d) Decommissioning phase

The institutions in charge of carrying out sound measurements must provide an accreditation according to DIN EN ISO/IEC 17025 or equivalent. The required certification must be presented on request to BSH.

Detailed information on the BSH measuring instruction is provided on ▶ www.bsh.de.

Literature

BSH (2013). Standard Investigation of the Impacts of Offshore Wind Turbines on the Marine Environment (StUK4). Bundesamt für Seeschifffahrt und Hydrographie, Hamburg and Rostock, 86 p.

BSH (2011). Offshore wind farms: Measuring instruction for underwater sound monitoring, Current approach with annotations. Bundesamt für Seeschifffahrt und Hydrographie, Hamburg and Rostock, 31 p.

Betke, K (2012). Underwater construction and operational noise at alpha ventus. RAVE (Research At *alpha ventus*) International Conference, May 8–10, 2012.

Betke K & Matuschek M (2012). Messungen von Unterwasserschall beim Betrieb der Windenergieanlagen im Offshore-Windpark *alpha ventus*. Oldenburg, May 2012 (report on behalf of the Federal Ministry for the Environment, Nature Conservation and Nuclear Safety (BMU) and Stiftung Offshore-Windenergie.

Betke K, Matuschek R, Nehls G, Grunau C (2012). Acoustical properties of an operational bubble curtain system. 11th European Conference on Underwater Acoustics, Edinburgh, July 2–6, 2012.

Betke K & Matuschek M (2011). Messungen von Unterwasserschall beim Bau der Windenergieanlagen im Offshore-Testfeld *alpha ventus*. Abschlussbericht zum Monitoring nach StUK3 in der Bauphase. Oldenburg, May 2011 (report on behalf of Stiftung Offshore-Windenergie).

Diederichs A, Brandt MJ, Nehls G, Laczny M, Hill A, Piper W (2010). Auswirkungen des Baus des Offshore-Testfelds *alpha ventus* auf marine Säugetiere. Husum, Juli 2010 (report on behalf of Stiftung Offshore-Windenergie).

Kastelein RA, Hoek L, de Jong CAF, Wensveen PJ (2010). The effect of signal duration on the underwater detection thresholds of a harbor porpoise (*Phocoena phocoena*) for single frequency-modulated tonal signals between 0.25 and 160 kHz. J. Acoust. Soc. Am., 128 (5).

Rustemeier J, Grießmann T, Rolfes R (2012a). Testing of bubble curtains to mitigate hydro sound levels at offshore construction sites (2007 to 2011). RAVE (Research At *alpha ventus*) International Conference, May 8–10, 2012.

Rustemeier J, Grießmann T, Betke K, Gabriel J, Neumann T, Küchenmeister M (2012b). Erforschung der Schallminderungsmaßnahme "Gestufter Blasenschleier (Little Bubble Curtain)" im Testfeld *alpha ventus* ('Schall *alpha ventus*'). Abschlussbericht zum gleichnamigen Forschungsvorhaben, Förderkennzeichen 0325122 A und 0325122B.

15

Noise mitigation systems and low-noise installation technologies

Tobias Verfuß

Federal Maritime and Hydrographic Agency,
Federal Ministry for the Environment, Nature Conservation and Nuclear Safety (Eds.)
Ecological Research at the Offshore Windfarm alpha ventus,
DOI 10.1007/978-3-658-02462-8_16, © Springer Fachmedien Wiesbaden 2014

16.1   Introduction

In line with Germany's national renewable energy targets, thousands of wind turbines are going to be installed in the North Sea and the Baltic in the coming decades. This expansion of offshore wind power is obliged to take place in a sustainable and environmentally sound manner.

There are three general foundation principles for offshore wind turbines. The substructure can (1) be anchored to the seabed with piles or suction buckets, (2) rest on the surface of the seabed, or (3) float in the water column. Which principle is suitable for a given windfarm site depends on water depth, soil properties, wind turbine design and installation method, and investment cost. Assuming (i) as the foundation principle of choice for the next decade, then impact pile driving will remain the main installation method.

Various studies have shown that pile driving can have a major impact on the subsea environment. Strong sound impulses can harm or even kill marine animals near the sound source, and may cause temporary displacement of sensitive species from their habitats. Since it takes thousands of impacts to drive one pile into the seabed, the underwater environment is set to be disturbed by millions of sound impulses (◘ Fig. 16.1). To mitigate this environmental impact, extensive research and development (R&D) effort has gone into the abatement of pile driving noise.

The R&D activities in this field relate to the development, full-scale testing and evaluation of noise mitigation measures and low-noise foundation installation technologies. A variety of methods and technical solutions have been qualified for the offshore wind sector in recent years. Relevant noise mitigation techniques are presented in Sect. 16.2. Most are commercially available or will be on the market soon. Low-noise foundation installation technologies comprise vibro piling, drilling, gravity-based foundations, suction buckets, and floating foundations. Some of these are expected to become available in coming years. Promising approaches are presented in Sect. 16.3.

In comparing the efficiency of noise mitigation measures and low-noise foundation installation technologies, it has to be considered that for impulsive sounds the sound exposure level (SEL) is the appropriate measure while for constant sounds it is the equivalent continuous sound level (Leq). Though both values vary in the underlying definitions, they can be regarded as comparable.

16.2   Noise mitigation measures

The portfolio of noise reduction techniques can be divided into primary and secondary mitigation measures. Primary measures comprise modification of the hydraulic hammer, adjustment of piling energy, prolongation of the ramming impulse by inserting a dynamic layer between hammer and pile, and modification of the pile itself. Secondary measures are characterised by an air barrier or sound-dampening obstacles placed between the pile and the water column. Popular approaches for secondary measures include big and small bubble curtains, the hydro sound damper approach, various kinds of casings, and cofferdams.

16.2.1   Primary mitigation measures

Adjustment of piling energy

One of the simplest primary noise mitigation measures is the adjustment of piling energy. Piling noise can be reduced by a certain amount by applying less energy to the hydraulic hammer. The correlation can be seen for the 'soft start' ramp-up procedure. There are some limitations, however. Depending on the soil properties, a certain energy level has to be applied for effective pile penetration. Lower energy levels may also mean that piling takes longer and causes greater temporary disturbance.

Impulse prolongation

Another primary mitigation measure is impulse prolongation. The aim here is to lengthen the contact time between hydraulic hammer and pile by inserting a dynamic layer or 'piling cushion'. This alters the shape of the sound impulse for a marked reduction in the resulting noise levels. As a rule of thumb, doubling the impulse length cuts noise by about 9 dB (Elmer et al. 2007a). Numerical simulations for the installation of a monopile with a diameter of 4.7 m into sandy soil show that inserting

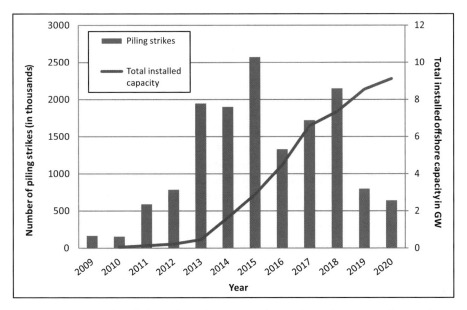

Fig. 16.1 Annual number of piling strikes in the German Bight according to the current and expected expansion of German offshore wind energy as of November 2013 (assuming 3,500 strikes per monopile, 10,500 strikes per tripile, 12,000 strikes per tripod, and 16,000 strikes per jacket foundation).

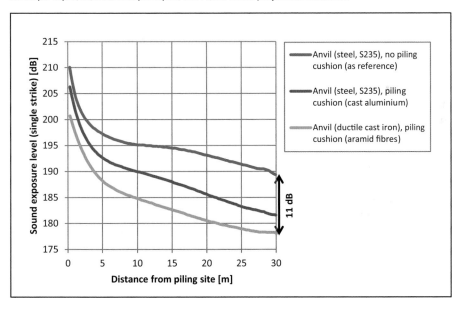

Fig. 16.2 Calculated decline of sound exposure level (SEL, single strike) with increasing distance from the piling site (after Rustemeier et al. (2012a)) for several hammer configurations with and without piling cushion.

a piling cushion made of aramid fibres can deliver noise reductions of up to 11 dB (SEL) or 13 dB (L_{peak}) (Rustemeier et al. 2012a; ◘ Fig. 16.2). Practical experience with the approach was gained during the installation of the substructure for the FINO2 research platform – a monopile with a diameter of 3.3 m – at Kriegers Flak in the Baltic Sea. The pile was driven into boulder clay beneath a sand layer. A coiled steel rope served as the piling cushion. A noise reduction of about 6 to 8 dB (SEL) was measured during the first few piling strokes. After that, however, the steel rope was too compressed to be effective (Elmer et al. 2007b). The theoretically derived noise reduction potential thus still has to be fully confirmed empirically. It must be noted that with high soil resistance, impulse prolongation can negatively affect pile drivability. This issue can be tackled by increasing the number of pile strikes or the ramming energy level.

16.2.2 Secondary mitigation measures

Bubble curtains

A bubble curtain is a layer of air bubbles over the full height of the water column. The bubbles are created by pumping compressed air through a perforated hose laid on the seabed or around a pile. Air bubble curtains were first used in connection with harbour works and bridge building. One of the first full scale tests was conducted in 2000 during the installation of bridge pillars for the San Francisco-Oakland Bay Bridge (Caltrans 2001).

To tailor the bubble curtain approach for offshore windfarm construction, it had to be adapted to greater water depths, the conditions of the harsh offshore environment, and serial use. It also had to be optimised for lean integration into the weather-dependent and time-critical windfarm installation process.

In principle, two bubble curtain layouts can be distinguished:
- 'Big' bubble curtains, which are laid at a radius from 70 to 150 m from a pile and enclose the whole construction site
- 'Small' bubble curtains, which are placed in the vicinity of a pile.

Big bubble curtains (BBCs)

A first full-scale test with a BBC was performed in June 2008 during installation of the substructure for the FINO3 research platform, a monopile with a diameter of 4.5 m (Grießmann et al. 2010). The non-optimised BBC prototype consisted of a hexagonal array of connected perforated plastic tubes at an average distance of 70 m around the construction site. This configuration made for a 12 dB reduction in sound exposure level (SEL) and a 14 dB reduction in peak level (L_{peak}). Based on the experience from this first test, an improved prototype BBC was developed and tested at the Trianel Windpark Borkum in winter 2011/2012 (Diederichs et al. 2014). The improved system consisted of a durable perforated hose attached to an anchor chain. The hose and chain were coiled on a drum on board an auxiliary vessel and deployed at the planned piling site before arrival of the jack-up barge. This meant that any delays in the windfarm installation process could be avoided. Several BBC setups, mainly differing in hose design (nozzle size), quantity of supplied air, and system layout (linear or circular), were successfully tested on 31 out of 40 tripod foundations. The most effective single configuration (◘ Fig. 16.3a), deployed in 12 instances, delivered an average noise reduction of 11 dB (SEL) and 14 dB (L_{peak}). Experiments with a double linear BBC indicated that greater reductions of up to 18 dB (SEL) and 22 dB (L_{peak}) may be feasible. The R&D project was the first ever serial test of a BBC during construction of an offshore windfarm. Similar BBC configurations have since been used at several other offshore windfarm projects in the German Bight. Experience has shown, however, that the attainable noise reduction can vary considerably. This depends on the BBC configuration, site-specific parameters (currents and seabed properties), and project-specific installation logistics requiring the BBC system to be laid at a certain minimum distance from the piling site.

Small bubble curtains (SBCs)

SBCs are positioned in the vicinity of a pile. This has the advantage compared with a BBC that no auxiliary vessel is needed to deploy the noise mitigation system.

A first full-scale test was conducted with a non-optimised prototype at *alpha ventus* in May 2009.

The SBC system consisted of two sections. The first comprised four perforated hoses mounted on the pile sleeves at the basis of the tripod foundation (◘ Fig. 16.3b). The second section, designed as a mobile system floating in the upper part of the water column, was intended to be connected to the fixed section. Attempts to connect the mobile system failed due to unfavourable weather conditions, so that only the lower part of the SBC could be deployed and put into operation. The noise measurements showed the efficiency of the system to heavily depend on the local current conditions. The sound attenuation measured against and with the direction of the current was 2/10 to13 dB (SEL) and 0/12 to 14 dB (L_{peak}) respectively (Rustemeier et al. 2012b), revealing an uneven formation of the bubble curtain and the presence of 'acoustic windows' facing against the current.

Learning from this experience, the SBC approach underwent a complete redesign. In 2012, an improved prototype was deployed and tested during installation of a tripile foundation at the BARD Offshore 1 windfarm (Steinhagen & Mesecke-Rischmann 2013). The SBC system (◘ Fig. 16.3c) comprised three hose arrays reeled out from drums at the top of the pile guidance frame. Each array consisted of eleven elastic polyethylene hoses with a length of 60 m, with one unperforated hose supplying the ten perforated hoses with air. So that different configurations could be tested, each array was able to be regulated separately. The air supply quantity could be varied over the test series between 0.07 and 0.13 m³/min·m. The best configuration (all three hose arrays active/maximum air supply) delivered a noise reduction of 13 dB (SEL) and 14 dB (L_{peak}) (Bellmann et al. 2014). The improved SBC system is commercially available. It has yet to undergo serial testing, however.

Hydro sound damper (HSD)

An HSD resembles a stationary bubble curtain placed in the vicinity of a pile (◘ Fig. 16.3d). In this approach, the air bubbles are replaced by air-filled balloons or robust foam elements of different sizes. These elements are arranged on a net that is kept in place with a ballast weight. Once lifted over the pile, the system can be released and recovered automatically. To enhance the sound damping ef-

ficiency, the number and composition of the HSD elements can be varied to suit the peak frequencies of the ramming noise. A test setup with a non-optimised prototype was investigated as part of the ESRa project (Wilke et al. 2012). This proved the functionality and applicability of the HSD concept. In August 2012, a full-scale test with an optimised HSD prototype was carried out at the London Array windfarm. The system was deployed during installation of one of the monopiles (◘ Fig. 16.3e). A broadband noise reduction of 7 to13 dB (SEL) and 7 to 15 dB (L_{peak}) was determined for a number of different measurement positions. Frequency analysis of the piling noise spectrum revealed a maximum damping efficiency of approximately 15 dB (SEL) between 200 and 500 Hz, while above 4 kHz no insertion loss could be recorded (Remmers & Bellmann 2013). The advantage of the HSD is that the system is independent of compressed air. The improved HSD is commercially available. However, it has not yet undergone serial testing.

Casings

A variety of casings have been proposed for the mitigation of piling noise. These can be pile sleeves of different materials or hollow steel tubes bigger in diameter than the pile. Casings are especially well suited for the installation of monopiles, but concepts for other foundation types are also under development. An example is the IHC Noise Mitigation System (NMS), a double-walled steel cylinder with sound-insulated connections between the inner and outer wall and an air-filled cavity (◘ Fig. 16.3f). A confined bubble curtain can also be activated inside the cylinder for greater efficiency. After a promising test of a prototype as part of the ESRa project, the NMS was technically scaled up and deployed at the Riffgat windfarm in 2012. All 30 monopile foundations were successfully installed using the noise mitigation system. Noise measurements were carried out for five pilings (van Vessem 2013). Lacking reference measurements, the achieved noise reduction can only be estimated. Taking the sound emissions measured at London Array (pile diameter 5.7 m) as a basis, the insertion loss for the NMS at Riffgat (pile diameter 5.7 to 6.5 m) was around 16 to 18 dB (SEL) and about 13 to 21 dB (L_{peak}). The use of casings is dependent on sufficient storage and crane

🔲 **Fig. 16.3** Secondary noise mitigation systems: (a) big bubble curtain (deployment at Trianel Windpark Borkum, November 2011), (b) small bubble curtain prototype (test at *alpha ventus*, May 2009), (c) improved small bubble curtain prototype on top of pile guidance frame (test at BARD Offshore 1, Sept. 2012), (d) HSD prototype for London Array, (e) HSD prototype (test at London Array, Aug. 2012), (f) casing NMS-6900 on-board of installation vessel Oleg Strashnov (deployment at Riffgat, Aug. 2012) (photo: (a) Hero Lang, (b-c) MENCK GmbH, (d-e) TU Braunschweig, (f) IHC MERWEDE).

capacity on the jack-up vessel. As the weight of a casing increases with water depth, the applicability of the method is limited.

Cofferdams

Significant sound attenuation can be achieved by executing offshore piling operations in air instead of in water. This principle forms the basis for cofferdams, which can be regarded as dewatered casings. According to Stokes et al. (2010), cofferdams have a mitigation potential of about 20 dB (SEL). This theoretical value was borne out by a demonstration project in the Danish Belt Sea (Århus Bugt) in December 2011. A cofferdam prototype was used on installation of a test pile with a diameter of 2.1 m at water depths of 14 to 24 m. After the cofferdam was placed onto the seabed and the pile inserted, the water was replaced with air using ejector pumps with a non-return valve. Piling thus took place in a dry environment. Accompanying noise measurements showed a noise reduction of 22 dB (SEL) and 18 dB (L_{peak}) (Thomsen 2012), underscoring the effectiveness of the cofferdam. However, a full scale test integrated into the windfarm installation process is still pending.

16.3 Low-noise foundation installation technologies

To date, most offshore wind turbines in the North Sea and Baltic have been mounted on piled foundations installed using hydraulic hammers. The dominant foundation type is the monopile, followed by jackets, tripods and triples. In some cases, vibratory piling has been used instead of impact pile driving. While gravity-based foundations have been deployed at some windfarm locations, mostly in shallower waters, little use has so far been made of other low-noise foundation installation technologies. Such technologies are heavily under development, however. Selected approaches are presented in the sections that follow.

16.3.1 Vibratory piling

One of the most common low-noise foundation installation technologies is vibratory piling or 'vibrodriving', where the hydraulic hammer is replaced by a vibratory hammer. This vibrates the pile in the vertical axis at a specific frequency. Vibratory piling is an option for installing piles in medium to dense bedded soils. While the method was previously limited to relatively small pile diameters, modern technology permits vibratory piling of monopiles with large diameters, currently of up to 6.5 m (de Neef et al. 2013). Calculations based on field measurements indicate that the resulting noise levels are 15 to 20 dB below the SEL associated with undamped impact pile driving (Elmer et al. 2007c), however on refusal or when driving through harder soil, noise

levels for vibrodriving may be higher. More recent experience shows such estimates to be somewhat conservative. Sound maxima in the emitted frequency spectrum are determined by peaks occurring at the vibration frequency (usually 20 to 40 Hz) and its harmonics.

Field experience with vibratory piling has been gained at various offshore windfarms, including *alpha ventus, Riffgat* and *Anholt*. At *alpha ventus*, the pin piles for the six tripod foundations were vibrated up to 9 m into the sandy seabed before applying hydraulic impact piling. The measured noise level (L_{eq}) was about 19 dB lower than with impact piling. Significantly lower noise sound pressure levels were apparent in the 100 Hz to 1 kHz frequency range (Betke & Matuschek 2011). At Riffgat, monopiles with diameters between 5.7 to 6.5 m were vibrated 15 to 30 m into the seabed with four vibro driving units before applying impact pile driving to bring the piles to target depth. The attainable penetration depth for each of the 30 monopiles varied with local soil properties. A noise level of 145 dB (L_{eq}) was measured for the vibratory piling (Fischer et al. 2013). This value is about 35 dB lower than the predicted SEL for Riffgat (Koschinski & Lüdemann 2013). However, it must be considered that vibratory piling at Riffgat was restricted to the upper part of the soil, and that the predicted SEL value is correlated to the highest expected noise immission levels at 750 m.

At Anholt, two monopiles with a diameter of 5.3 m were installed in sandy soil using three vibro driving units (Kringelum 2013, pers. comm.). Both piles were vibrated 17 to 18 m into the seabed. One of the piles met refusal at this depth due to very dense sand and had to be driven by impact pile driving for the last metres. In the other case impact driving was used to obtain some reference noise measurements. L_{eq} recorded for vibratory piling was about 15 to 17 dB below the SEL measured during impact piling. The peak level was reduced by 27 dB (L_{peak}). During vibrodriving in dense sand the noise levels increase and the noise reduction compared to impact driving is significantly less.

From the experience gained, it can be concluded that vibratory piling is well suited for the installation of monopile foundations in various soil conditions including both medium to densely bedded sands and cohesive soils. Compared with the SEL associated with undamped impact pile driving, the continuous sound level (L_{eq}) for vibratory piling is about 15 to 30 dB lower. However, vibratory piling is currently restricted to the upper metres, as current design guidelines for proof of axial capacity are limited to impact driven piles. In general, the proof for axial loading of monopiles is not considered critical as the design is governed by the lateral loading.

16.3.2 Offshore foundation drilling (OFD)

As an alternative to impact pile driving, piled foundations can be installed by drilling. This method is especially well suited for 'difficult' soil conditions (e.g. rocky seabed) and in principle can be used at water depths of up to 80 m. Due to higher cost relative to other installation methods, however, drilling technology has so far seen only occasional use. At the UK Barrow windfarm, three monopiles with a diameter of 4.75 m were drilled using the BFD 5500 Flydrill system into a seabed consisting of various layers of dense sand, silt, clay, mudstone/siltstone and siltstone/sandstone. Progress rates varied between 0.35 and 1.0 m per hour (Beyer & Brunner 2006). Unfortunately, no sound recordings are available for the drilling operations. More experience was gained in 2008, when a 5 m wide shaft was drilled 39 m into the seabed at an onshore location in Naples, Italy, with a vertical shaft machine. As the drilling partly took place in a water-saturated environment, the drilling noise could be monitored with hydrophones. Transferring the measured noise levels to offshore conditions yields a sound pressure level (L_{eq}) of 117 dB and a peak sound pressure level (L_{peak}) of 122 dB at a distance of 750 m from the noise source (Rustemeier et al. 2012a). These values are rough estimates and should be treated with care; however, they indicate that drilling can be regarded as a low-noise foundation installation technology.

To adapt the technology for the serial installation of wind turbine substructures, several approaches are presently under further development. These include the Drilled Concrete Monopile approach and Offshore Foundation Drilling (OFD) (Koschinski & Lüdemann 2013). In the latter, a

drilling machine with a cutterhead is clamped inside the pile and cuts its way vertically through the seabed regardless of the soil properties. An improved version of the approach is expected to reach cutting speeds of 3.0 m per hour (Jung 2013).

In its current state, drilling technology is not ready for standard use. Efforts focus on reducing the cost by enhancing the approach with higher cutting speeds and greater tolerance to offshore weather conditions, and on allowing the installation of large concrete monopiles with diameters upwards of 10 m and massive wall thicknesses.

16.3.3 Suction buckets / suction cans

Instead of using piles, offshore structures can be anchored in the seabed with suction buckets or suction cans. This approach has proven to be suitable for auxiliary platforms used by the oil and gas industry, and also for transformer substations and met masts (Offshore Engineer 2005, GlobalTech1 2013, Universal Foundation 2013). Although initial experience with the erection of wind turbines on bucket foundations was gathered about a decade ago (Ibsen et al. 2005), the approach does yet not play any part in the installation of windfarms due to the dynamic loads created by the operation of offshore wind turbines. The ensuing risks for the stability of wind turbine substructures are yet to be sufficiently assessed.

The suction bucket principle is a low-noise foundation installation technology, as the only noise source is the suction pump used to evacuate water and loose sediment from the interior of the suction bucket/can. The installation process can be described as follows: First, the suction bucket is placed onto the seabed. Due to the deadweight of the substructure, the bucket penetrates the soil to a certain depth. Subsequently, negative pressure is applied to the interior of the bucket by means of several suction pumps. With ongoing evacuation, the hydrostatic pressure difference and the deadweight of the substructure cause the bucket to penetrate the soil. After installation is complete, the foundation acts like a hybrid of a gravity-based structure and a monopile.

Depending on the substructure design, suction foundations can consist of one big suction bucket

Fig. 16.4 Design for suction bucket jacket foundation prototype (photo: DONG Energy).

('monopod') or several smaller suction cans ('multipod'). Monopod buckets have typical diameters of 9 to 15 m. They consist of several chambers that can be evacuated separately for precise levelling of the foundation. Multipod designs are suitable for the installation of tripods or jackets (**Fig. 16.4**). In this case, smaller suction cans with a diameter of 3 to 6 m are placed at each corner of the substructure. Levelling is accomplished by pumping operations alternating between the suction cans.

Suction bucket foundations can be applied in both deep and shallow waters. However, it must be noted that their use is restricted to sandy soils, clay and other homogenous water-saturated sediments.

16.4 Perspectives

To date, impact pile driving has been the predominant method of installing substructures for offshore wind turbines. Germany has adopted strict regulations on impulsive noise (such as pile driving noise), making the use of noise mitigation technologies mandatory (▶ see Chap. 1, ▶ Information box in Chap. 13). Alternatively, the offshore wind industry can deploy low-noise foundation installation technologies.

In recent years, various noise mitigation measures and low-noise foundation installation technologies have been refined and adapted for integration into windfarm installation processes. A variety of secondary measures are commercially available, as seen in the comparative overview in **Table 16.1**. Big bubble curtains and tubular steel casings have been successfully applied in windfarm projects and can thus be regarded as state of the art. Other approaches still remain at the prototype or concept stage. Vibratory piling has proven to be suitable for the installation of monopile foundations in sandy soils.

As demonstrated at the Trianel Windpark Borkum and the Riffgat windfarm, piling noise can be kept below the prescribed thresholds for sound exposure level and peak level in certain circumstances (Diederichs et al. 2014, van Vessem 2013). However, no technology can yet assure full compliance with the noise threshold criteria. Especially with regard to 'difficult' soil conditions and the installation of large diameter piles for multi-megawatt offshore turbines, further effort is necessary to enhance the effectiveness of existing noise mitigation technologies, to develop novel, more effective solutions, or to switch to low-noise foundation installation technologies. To cut costs, noise mitigation should be not regarded as an 'add-on' requirement, but should already be considered during the design phase of future windfarm projects.

◼ **Table 16.1** State of technology (as of November 2013) for selected secondary noise mitigation measures and low noise foundation installation technologies.

	Feasibiliy study completed	Prototype test completed	Demonstration during OWF construction	Serial use demonstrated	Commercially available
Big bubble curtain	✓	✓	✓	✓	✓
Small bubble curtain	✓	✓	✓		✓
Hydro sound damper	✓	✓	✓	Expected 2014	✓
Casing (Noise Mitigation Screen)	✓	✓	✓	✓	✓
Cofferdam	✓	✓			✓
Vibratory piling	✓	✓	✓*	✓*	✓
Offshore Foundation Drilling	✓				
Suction Bucket Jacket Foundation	✓		Expected 2014		

* yet not applicable for reaching target depth

Literature

Bellmann M, Gündert S, Remmers P (2014). Offshore Messkampagne 1 (OMK 1) für das Projekt BORA im Windpark BARD Offshore 1. Report for the BMU funded research project "Predicting Underwater Noise due to Offshore Pile Driving (BORA)", project ref. no. 0325421 A. Oldenburg, February 2014, 141 pp.

Betke K & Matuschek R (2011). Messungen von Unterwasserschall beim Bau der Windenergieanlagen im Offshore-Testfeld *alpha ventus*. Abschlussbericht zum Monitoring nach StUK3 in der Bauphase. Oldenburg, May 2011, 48 pp.

Beyer M & Brunner WG (2006). Drilling Monopiles at *Barrow* Offshore Windfarm, UK. Tiefbau 6/2006, p. 346–350.

Caltrans (2001). Marine Mammal Impact Assessment for the San Francisco – Pile Installation Demonstration Project. Report for Oakland Bay Bridge East Span Seismic Safety Project. California Department of Transportation (Caltrans). Contract No. 04A0148, 04-ALA-80-0.0/0.5, August 2001, 49 pp.

de Neef L, Middendorp P, Bakker J (2013). Installation of Monopiles by Vibrohammers for the Riffgat Project. In: Proceedings Pfahlsymposium 2013, Braunschweig, 21 February 2013, 14 pp.

Diederichs A, Pehlke H, Bellmann M, Gerke P, Oldeland J, Grunau C, Witte S, Rose A, Nehls G (2014). Entwicklung und Erprobung des Großen Blasenschleiers zur Minderung der Hydroschallemissionen bei Offshore-Rammarbeiten. Final report for the BMU funded research project "Hydroschall-OFF BWII", project ref. no. 0325309 A/B/C. Husum, March 2014, 240 pp.

Elmer KH, Betke K, Neumann T (2007a). Standardverfahren zur Ermittlung und Bewertung der Belastung der Meeresumwelt durch die Schallimmission von Offshore-Windenergieanlagen. Final report for the BMU funded research project "Schall II", project ref. no. 0329947. Hanover, March 2007, 129 pp.

Elmer KH, Betke K, Neumann T (2007b). Standardverfahren zur Ermittlung und Bewertung der Belastung der Meeresumwelt durch die Schallimmission von Offshore-Windenergieanlagen – Untersuchung von Schallminderungsmaßnahmen an FINO 2. Final report for the correspondent BMU funded research project, project ref. no. 0329947 A. Hanover, July 2007, 28 pp.

Elmer KH, Gerasch WJ, Neumann T, Gabriel J, Betke K, Schultz-von Glahn M (2007c). Measurement and Reduction of Offshore Wind Turbine Construction Noise. DEWI Magazine No. 30. Wilhelmshaven, February 2007, p 33–38.

Fischer J, Sychla H, Bakker J, de Neef L, Stahlmann J (2013). A comparison between impact driven and vibratory driven steel piles in the German North Sea. In: Proceedings Conference on Maritime Energy (COME), Hamburg, 21–22 May 2013.

GlobalTech1 (2013). Erfolgreiche Installation der parkinternen Umspannstation. Press release 09.05.2013.

Grießmann T, Rustemeier J, Betke K, Gabriel J, Neumann T, Nehls G, Brandt M, Diederichs A, Bachmann J (2010). Erforschung und Anwendung von Schallminimierungsmaßnahmen beim Rammen des FINO3-Monopiles. Final report for the BMU funded research project "Schall FINO3", project ref. no. 0325023 A/0325077. Hanover 2010, 144 pp.

Ibsen LB, Liingaard M, Nielsen SA (2005). Bucket foundation, a status. In: Proceedings of Copenhagen Offshore Wind 2005. Copenhagen, 26–28 October 2005.

Jung B (2013). Marktreife Weiterentwicklung der VSM-Technologie für die Erstellung von Fundamenten für Offshore-Windenergieanlagen. Final report for the correspondent BMU funded research project, project ref. no. 0325233. Schwanau, March 2013, 32 pp.

Koschinski S & Lüdemann K (2013). Entwicklung schallmindernder Maßnahmen beim Bau von Offshore-Windenergieanlagen. Studie im Auftrag des Bundesamtes für Naturschutz (BfN). Nehmten/Hamburg, February 2013, 96 pp.

Kringelum JV (2013, pers. comm.). Results of Anholt noise mitigation project by DONG energy. To be presented at EWEA Offshore 2013 conference.

Offshore Engineer 2005: MOAB scores North Sea first. Offshore Engineer, Nov. 2005, p. 25–28.

Remmers P & Bellmann M (2013). Untersuchung und Erprobung von Hydroschalldämpfern (HSD) zur Minderung von Unterwasserschall bei Rammarbeiten für Gründungen von Offshore-Windenergieanlagen. Report for the BMU funded research project "HSD", project ref. no. 0325365. Oldenburg, June 2013, 55 pp.

Rustemeier J, Neuber M, Grießmann T, Ewaldt A, Uhl A, Schultz-von Glahn M, Betke K, Matuschek R, Lübben A (2012a). Konzeption, Erprobung, Realisierung und Überprüfung von lärmarmen Bauverfahren und Lärmminderungsmaßnahmen bei der Gründung von Offshore-WEA. Final report for the BMU funded research project "Schall III", project ref. no. 0327645. Hanover, May 2012, 236 pp.

Rustemeier J, Grießmann T, Betke K, Gabriel J, Neumann T, Küchenmeister M (2012b). Erforschung der Schallminderungsmaßnahme "Gestufter Blasenschleier" (Little Bubble Curtain) im Testfeld *alpha ventus*. Final report for the BMU funded research project "Schall *alpha ventus*", project ref. no. 0325122 A/B. Hanover/Kaltenkirchen, July 2012, 79 pp.

Steinhagen U & Mesecke-Rischmann S (2013). Untersuchung und Erprobung eines 'Kleinen Blasenschleiers' zur Minderung von Unterwasserschall bei Rammarbeiten für Gründungen von OWEA. Final report for the BMU funded research project "Hydroschall-OFF BO1", project ref. no. 0325334G. Kaltenkirchen, July 2013, 48 pp.

Stokes A, Cockrell K, Wilson J, Davis D, Warwick D (2010). Mitigation of Underwater Pile Driving Noise During Offshore Construction. Report no. M09PC00019 to Department of the Interior, Minerals Management Service. Groton/CT, January 2010, 104 pp.

Thomsen KE (2012). Kofferdam test piling in Århus – Including the sound mitigation concept for Borwin2 and Helwin1. Århus, January 2012, 17 pp.

Universal Foundation (2013). Universal Foundation Suction Bucket – A solution in support for offshore wind. Presentation Henrik Lundorf Nielsen, April 2013.

van Vessem H, van Erkel T, Jung B (2013). IHC Hydrohammer Noise Mitigation Screen: history, improvements, back-up systems, research for XL-piles and new awarded projects.

Presentation at the StUKplus conference. Berlin, 31st October 2013.

Wilke F, Kloske K, Bellmann M (2012). Evaluation of Systems for Ramming Noise Mitigation at an Offshore Test Pile. Final report for the BMU funded research project "ESRa", project ref. no. 0325307. Hamburg, May 2012, 168 pp.

Cumulative impacts of offshore windfarms

Hans-Peter Damian & Thomas Merck

Federal Maritime and Hydrographic Agency,
Federal Ministry for the Environment, Nature Conservation and Nuclear Safety (Eds.)
Ecological Research at the Offshore Windfarm alpha ventus,
DOI 10.1007/978-3-658-02462-8_17, © Springer Fachmedien Wiesbaden 2014

17.1   Introduction

The use of wind energy in marine areas only started a couple of decades ago. Offshore windfarms intrude upon an environment already heavily affected by human activities such as shipping and fishing. As they are large and complex installations and given the large numbers planned – at least in European waters – offshore windfarms can be expected to have significant impacts on marine ecosystems with long-lasting effects from both construction and operation.

Environmental impacts of offshore windfarms include destruction of the sea bottom and benthic communities, disruption of migrating species such as birds via barrier effects, and disturbance of sound-sensitive marine species through increased underwater noise. The various impacts affect species both concurrently and sequentially at different life stages and should therefore be assessed from a cumulative perspective.

Providing an environmentally suitable location has been chosen, a single windfarm does not necessarily have much more than a localised impact on the marine environment. Highly mobile species like marine fish, birds and mammals, however, will encounter offshore windfarms repeatedly in the course of a year. Multiple offshore windfarms can therefore have cumulative negative impacts on the conservation status of such species. For this reason we will focus on seabirds and migrating birds, and on underwater noise and its impact on marine mammals.

17.2   Seabirds

Seabirds show a range of different behaviours when confronted with offshore wind turbines. Some species do not hesitate to fly into windfarms to forage (some gull and tern species) or even to use the structures for resting (cormorant) (Dierschke & Garthe 2006). Birds of these species generally face a risk of collision with turbines (Everaert & Stienen 2007). Other species avoid windfarms either partly, leading to reduced bird densities (e.g. long-tailed duck) (Petersen et al. 2011), or almost entirely (e.g. red-throated diver and black-throated

diver; gannet) (Dierschke & Garthe 2006), resulting in habitat loss.

A number of species are displaced by moving ships (Schwemmer et al. 2011), with service boats on the way to and from windfarms causing disturbance and temporary habitat loss. Indirect impacts can arise during windfarm construction, when pile driving displaces fish and thus reduces the food supply for fish-eating seabirds (Perrow et al. 2011). On the other hand, food supply may increase in the operational phase, because fisheries are excluded from windfarms and epibenthic organisms settle on the introduced structures.

In the case of lethal collisions, the effects of multiple windfarms can simply be added up, because the effects of one windfarm do not influence those of others. When it comes to habitat loss from avoidance and disturbance, too, the effects of multiple windfarms likewise at first appear simply to add up, with each additional windfarm increasing the impact on the various seabird populations. As avoidance includes the effect on birds flying through windfarms (Aumüller et al. 2013) and thus detours (Aumüller et al. 2013, Krijgsveld et al. 2010), fragmentation of the available habitat also has to be considered, resulting in greater habitat loss as modelled by Busch et al. (2013). Hence in a cumulative assessment of offshore windfarms the effects may act synergistically – all the more so since potential density effects in alternative marine habitats have not been investigated.

Future offshore windfarms should therefore be planned away from important seabird habitats to avoid high rates of collisions and habitat loss. Consideration should also be given to leaving corridors free of windfarms between seabird habitats so that birds can safely switch between sites.

17.3   Migrating birds

For various species of migrating birds, offshore windfarms act as barriers in daytime while lethal collisions predominantly take place at night. The species spectra affected, however, differ fundamentally depending on whether species migrate diurnally or nocturnally.

Many species give windfarms a wide berth, including ducks, geese, swans, waders and auks plus certain other species such as fulmar, gannet, little gull, kittiwake and sandwich tern. Large gulls of the genus Larus, black-headed gull and common gull, in contrast, do not seem to be affected by windfarms, whereas it is possible that passerine species migrating during the day may even be attracted.

Reactions towards windfarms clearly differ between species. Severe consequences resulting from deviations in route could arise from future additions to the number of windfarms. Because of the additional energy demands, longer migration routes can affect both immediate survival and subsequent breeding success.

The literature suggests that none of the barrier effects identified so far have significant impacts on populations. However, there are circumstances where a barrier effect might lead indirectly to population-level impacts, for example where a windfarm effectively blocks a regularly used flight line between nesting and foraging areas, or where several windfarms cumulatively interact to create an extensive barrier that could result in diversions of many tens of kilometres (Drewitt et al. 2006).

Nocturnal migration is dominated by passerine species, in particular thrushes. Avian casualties at research platforms reflect the nocturnal species spectrum (Hüppop et al. 2012). Collisions with anthropogenic structures occur most of all when good weather conditions for migration worsen (e.g. clear sky changing to fog and drizzle, tailwinds turning into headwinds). In such conditions, birds tend to head for light sources such as those found on platforms and wind turbines.

Although birds generally migrate at lower heights over sea than over land, good weather conditions allow for migration at heights that do not pose risks for collision and where light sources are widely ignored (Hüppop et al 2006).

Migrations encountering anthropogenic structures at sea can lead to hundreds of casualties in single night, as seen for the research platform FINO1 (Hüppop et al. 2006, for causes see also Aumüller et al. 2011). Even though they seem impressive, however, the numbers remain flawed: It has not yet been possible to quantify the mortality rate in relation to actual migration intensity. The increase in mortality is probably well within the capacity of a population to compensate for additional losses

(to regenerate) and hence has no effect on overall population levels. However, the cumulative increase in mortality resulting from a number of windfarms may exceed a population's capacity for regeneration.

Even state-of-art methodologies are unable to quantify nocturnal migration at species level. Impact assessments at population level are therefore limited.

To protect migratory birds, it is recommended to provide lighting for offshore wind energy turbines, in line with demand. In nights of high migration when the weather is bad and visibility is poor, the approval authority (BSH) reserves the right to have the turbines temporarily switched off after having evaluated the situation (▶ Information box *The Incidental Provision 21*, see Chap. 12). At the spatial planning stage, it is crucial to avoid dead-end corridors between windfarms. The effects of corridor widths on migrating birds also need to be evaluated.

17.4 Underwater noise and marine mammals

Underwater sound naturally plays an important role for a number of aquatic or marine animals. Marine mammals such as whales in particular use natural and self-generated sounds to navigate, to detect food or predators, and to communicate with each other (Richardson et al. 1995). Anthropogenic noise entering the marine environment has the potential to impair these biologically important functions.

Underwater noise has become an issue of major concern in recent years with regard to human impacts on the marine environment in general (e.g. OSPAR 2009). In some marine areas, for example, the level of ambient noise has doubled every decade for the last 35 years (McDonald et al. 2006, Andrew et al. 2011). Noise from offshore windfarms and from their installation thus enters an already noisy environment and has to be assessed cumulatively.

Ambient noise demonstrates the fundamental cumulative character of underwater noise (◻ Fig. 17.1). Beside natural sounds generated by wind, waves and rain as well as various biological sounds, shipping contributes the most to ambient noise. In addition, there are localised but extremely loud anthropogenic sound events like seismic surveys, explosions and pile driving.

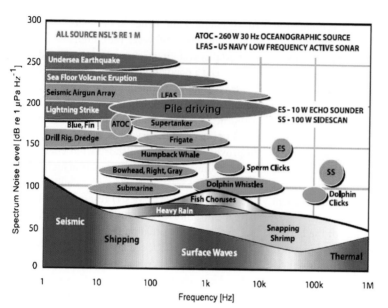

Fig. 17.1 Underwater noise of different sources. The frequency range and noise level of natural and anthropogenic underwater sounds are shown (Coates 2002, modified).

Regarding noise emissions from offshore windfarms, major concerns have been raised about the impact of pile driving especially on marine mammals. Steel tubes measuring several meters in diameter are driven by impact hammering into the sea bottom to form or fix the foundations of offshore wind turbines. Impulsive noise from pile driving has the potential to disturb or even injure marine mammals (e.g. Tougaard et al. 2006, OSPAR 2006, 2009).

The most common whale species in German waters is the harbour porpoise. Like other cetaceans, the harbour porpoise strongly depends on its sense of hearing for navigation, feeding and communication. Injuries in terms of hearing impairments such as a temporal threshold shift (TTS) occur at sound levels well below the source level emitted by a driven pile (Lucke et al. 2009, OSPAR 2009). In addition, as observed during construction of the Danish Horns Rev 1 windfarm, noise-induced disturbance of harbour porpoises can reach distances of more than 25 km from the pile driving site, thus covering an area of about 2,000 km² (Tougaard et al. 2006). It should be mentioned that in the German Exclusive Economic Zone (EEZ), noise mitigation measures have to be applied when pile driving and emissions may not exceed a certain level to ensure that no harbour porpoises suffer TTS. A number of efficient noise mitigation measures and even low-noise foundations are available or under development (Koschinski & Luedemann 2013, ► see Chap. 16).

Neglecting other anthropogenic sound sources and considering only offshore windfarms due to the large number of projects, cumulative impacts occur on both a spatial and a temporal scale. In the German North Sea, more than 100 projects with up to 80 turbines each have been applied for, of which 30 have already been approved or are under construction or operational (■ Fig. 17.2).

It can be projected that a number of projects will enter the construction phase each year for the next 20 to 30 years, emitting a large amount of anthropogenic impulsive noise into the marine environment. While compliance with the German noise threshold will substantially reduce the disturbance radius, the area affected will still add up to some hundreds of square kilometres, and the level of disturbance will be sustained year in, year out. Disturbance in the biological sense includes changes in behaviour, lost feeding time, or expenditure of extra energy to escape the area.

If multiple windfarms are built at the same time with pile driving done in alternation, animals escaping one disturbed area may enter another where pile driving then starts up, thus again suffering noise-induced disturbance. Such energy-related aspects are of special relevance during the most sensitive

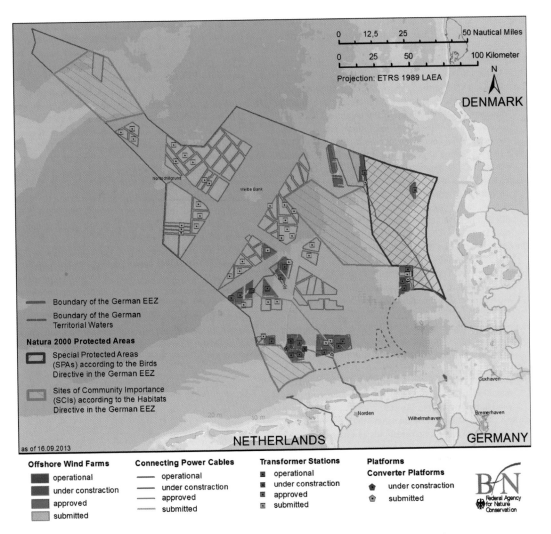

Fig. 17.2 State of development of offshore windfarms in the German North Sea (as of September 2013).

time of the reproduction phase when harbour porpoises give birth to and nurse calves.

On the other hand, cumulative impacts can even occur at the level of a single windfarm. Driving a pile into the sea bottom takes hundreds or thousands of hammer strokes. The cumulative energy impact on the ear of a receiving animal can induce TTS even if the sound level emitted by a single stroke does not have the capacity to do so (Southall et al. 2007).

In conclusion, windfarms will have cumulative impacts at various scales and affecting various features of marine life. Such impacts can be reduced or avoided by careful siting and by coordinating windfarm construction in both time and space.

Literature

Aumüller R, Boos K, Freienstein S, Hil, K, Hill R (2011). Beschreibung eines Vogelschlagereignisses und seiner Ursachen an einer Forschungsplattform in der Deutschen Bucht – Vogelwarte 49: 9–16.

Aumüller R, Boos K, Freienstein S, Hill K, Hill R (2013). Weichen Zugvögel Windenergieanlagen auf See aus? Eine Methode zur Untersuchung und Analyse von Reaktionen tagsüber ziehender Vogelarten auf Offshore-Windparks. – Vogelwarte 51: 3–13.

Andrew RK, Howe BM, Mercer JA (2011). Long-time trends in ship traffic noise for four sites off the North American West Coast. – J. Acoust. Soc. Am. 129, 642–651.

Coates R (2002). The Advanced SONAR Course. – Seiche ISBN 1-904055-01-X.

Busch M, Kannen A, Garthe S, Jessop M (2013). Consequences of a cumulative perspective on marine environmental impacts: Offshore windfarming and seabirds at North Sea scale in context of the EU Marine Strategy Framework Directive. Ocean & Coastal Management 71 (2013) 213–224.

Dierschke V & Garthe S (2006). Literature review of offshore windfarms with regard to seabirds. – In: Zucco C, Wende W, Merck T, Köchling I, Köppel J (eds.): Ecological Research on Offshore Windfarms: International Exchange of Experiences. Part B: Literature Review of Ecological Impacts. BfN-Skripten 186, Bundesamt für Naturschutz, Bonn, pp. 131–198.

Drewitt AL & Langston RHW (2006). Assessing the impacts of windfarms on birds. – Ibis 148: 29–42.

Everaert J & Stienen EWM (2007). Impact of wind turbines on birds in Zeebrugge (Belgium) – significant effect on breeding tern colony due to collisions. – Biodivers. Conserv. 16: 3345–3359.

Hüppop O & Hilgerloh G (2012). Flight call rates of migrating thrushes: effects of wind conditions, humidity and time of day at an illuminated offshore platform. – J. Avian Biol.: 85–90.

Hüppop O, Dierschke J, Exo KM, Fredrich E, Hill R (2006). Bird migration studies and potential collision risk with offshore wind turbines – Ibis 148: 90–109.

Lucke K, Siebert U, Lepper PA, Blanchet MA (2009). Temporary shift in masked hearing thresholds in a harbor porpoise (Phocoena phocoena) after exposure to seismic airgun stimuli. – J. Acoust. Soc. Am. 125: 4060–4070.

Krijgsveld KL, Fijn RC, Heunks C, van Horssen PW, de Fouw J, Collier MP, Poot MJM, Beuker D, Dirksen S (2010). Effect Studies Offshore Windfarm Egmond aan Zee. Progress report on fluxes and behaviour of flying birds 33 covering 2007 & 2008. Bureau Waardenburg report 09-023. Bureau Waardenburg, Culemborg.

Koschinski S & Luedemann K (2013). Development of Noise Mitigation Measures in Offshore Windfarm Construction. – Report to the German Federal Agency for Nature Conservation, BfN, Bonn.

McDonald MA, Hildebrand JA, Wiggins SM (2006). Increases in deep ocean ambient noise in the Northeast Pacific west of San Nicolas Island, California. – J. Acoust. Am. 120(2):711–718.

OSPAR (2006). Review of the Current State of Knowledge on the Environmental Impacts of the Location, Operation and Removal/Disposal of Offshore Wind-Farms. – OSPAR Commission, 2009. Publication number 278/2006.

OSPAR (2009). Overview of the impacts of anthropogenic underwater sound in the marine environment. – OSPAR Commission, 2009. Publication number 441/2009.

Petersen IK, MacKenzie M, Rexstad E, Wisz MS, Fox AD (2011). Comparing pre- and post-construction distributions of long-tailed ducks Clangula hyemalis in and around the Nysted offshore windfarm, Denmark: a quasi-designed experiment accounting for imperfect detection, local surface features and autocorrelation. – CREEM Tech Report 2011-1.

Perrow MR, Gilroy JJ, Skeate ER, Tomlinson ML (2011). Effects of the construction of Scroby Sands offshore windfarm on the prey base of Little tern Sternula albifrons at its most important UK colony. – Mar. Poll. Bull. 62: 1661–1670.

Richardson WJ, Malme CI, Green Jr CR, Thomson DH (1995). Marine mammals and noise. – Vol 1. Academic Press, San Diego, California, USA.

Schwemmer P, Mendel B, Sonntag N, Dierschke V, Garthe S (2011). Effects of ship traffic on seabirds in offshore waters: implications for marine conservation and spatial planning. – Ecol. Appl. 21: 1851–1860.

Southall BL, Bowles AE, Ellison WT, Finneran JJ, Gentry RL, Greene jr. CR, Kastak D, Ketten DR, Miller JH, Nachtigall PE, Richardson WJ, Thomas JA, Tyack PL (2007). Marine mammal noise exposure criteria: Initial scientific recommendations. – Aquat. Mamm. 33, 411–414.

Tougaard J, Carstensen J, Wisz MS, Jespersen M, Teilmann J, Bech NI, Skov H (2006). Harbour Porpoises on Horns Reef Effects of the Horns Reef Windfarm. – Final Report to Vattenfall A/S, NERI, Roskilde, Denmark.

Backmatter

Links

Federal Maritime and Hydrographic Agency,
Federal Ministry for the Environment, Nature Conservation and Nuclear Safety (Eds.)
Ecological Research at the Offshore Windfarm alpha ventus,
DOI 10.1007/978-3-658-02462-8, © Springer Fachmedien Wiesbaden 2014

Links

---

alpha ventus offshore windfarm
www.alpha-ventus.de

Agreement on the Conservation of African-Eurasian Migratory Waterbirds (AEWA)
www.unep-aewa.org

ACCOBAMS
www.accobams.org

ASCOBANS
www.ascobans.org

Birdlife International
www.birdlife.org

BSH Standards
www.bsh.de/en/Products/Books/Standard/

Convention on Migratory Species (CMS)
www.cms.int/documents/convtxt/cms_convtxt_english.pdf

Ecological monitoring in accordance with StUK3
www.bsh.de/en/Marine_uses/Industry/Wind_farms/Ecological_monitoring.jsp

Ecological research accompanying the *alpha ventus* offshore test project
www.bsh.de/de/Meeresnutzung/Wirtschaft/Windparks/StUKplus/stukplustext.jsp

Federal Agency for Nature Conservation/Habitatmare
www.habitatmare.de

Federal Environment Agency
www.uba.de

Federal Maritime and Hydrographic Agency
www.bsh.de

Federal Ministry for the Environment, Nature Conservation, Building and Nuclear Safety
www.bmub.bund.de

German Energy Agency offshore wind power portal
www.offshore-wind.de

Natura 2000 in the EU
http://ec.europa.eu/environment/nature/index_en.htm

North Sea and Baltic Sea Monitoring Programme
www.blmp-online.de/

Helsinki Commission
www.helcom.fi/

Oslo-Paris Commission
www.ospar.org/

Projekt Management Jülich
www.ptj.de/en/start

RAVE Research Initiative
http://rave.iwes.fraunhofer.de/rave/pages/welcome

Seabirds at Sea – Europe
www.jncc.gov.uk/page-1547

Seabirds at Sea – Germany
www.uni-kiel.de/ftzwest/ag4/projekte/birds/sas-e.shtml

StUKplus Project
www.stukplus.com

Printing: Ten Brink, Meppel, The Netherlands
Binding: Stürtz, Würzburg, Germany